# HELPING NATURE HEAL

## AN INTRODUCTION TO ENVIRONMENTAL RESTORATION

Edited by Richard Nilsen

Foreword by Barry Lopez

A WHOLE EARTH CATALOG/TEN SPEED PRESS PUBLICATION

## Helping Nature Heal

*Editor:* Richard Nilsen
*Designer:* Kathleen O'Neill
*Production Managers:* Jonathan Evelegh,
Lori Woolpert
*Copy Editor, Typesetter, Pasteup:*
James Donnelly
*Camera, Pasteup, Design:* Donald Ryan
*Pasteup:* Charlie Oldham
*Proofreader:* Hank Roberts
*Index, Bibliography:* David Burnor
*Editorial Assistant:* Corinne Cullen Hawkins
*Project Manager:* Keith Jordan

*Point Foundation Staff:*
Kelly Teevan (Executive Director), J. Baldwin,
Christine Goodson, Kurt Grubaugh,
Richard Kadrey, Kevin Kelly, Peter Klehm,
Jason Mongue, Howard Rheingold,
Susan Rosberg, Sarah Satterlee

*Point Foundation Board:* Stewart Brand,
Doug Carlston, Christina Desser,
Robert Fuller, Huey Johnson,
Peggy Lauer, Kelly Teevan

*Cover Design:* Nancy Austin
*Front cover photos (left to right):*
Charles Cadmon, William Campbell,
Salt Force News, William Campbell.
*Back cover photo:* Salt Force News.

**For information on ordering books
whose access information includes
the words ''or Whole Earth Access,''
see page 152.**

**This book had its origins as a
special issue of Whole Earth
Review. For information on how
to subscribe to Whole Earth
Review, see page 152.**

**Whole Earth Catalog/Ten Speed Press Books**
are the collaborative endeavor of the makers of
the Whole Earth Catalog series and Ten Speed
Press. Our hope is to expand the power of individuals and communities to conduct their own
education, find their own inspiration, and
shape their own environment.

This book is printed
on recycled paper that
meets EPA standards.

Copyright © 1991 by Point Foundation

Ten Speed Press
Box 7123
Berkeley, CA 94707

**Library of Congress Cataloging-in-Publication Data**

Helping nature heal: an introduction to environmental
restoration/edited by Richard Nilsen; foreword by
Barry Lopez.
    p. cm.
    ''A Whole Earth Catalog.'' Includes bibliographical
references and index.
    ISBN 0-89815-425-1
    1. Restoration ecology.  2. Restoration ecology —
United States.
I.  Nilsen, Richard.
QH541.15.R45A45  1991           91-4750
333.7′ 153 — dc20                  CIP

Printed in the United States of America

1  2  3  4  5  -  94  93  92  91

# CONTENTS

# FOREWORD

The Earthly ills are so apparent and the lack of courage in government and industry, the usual whipping boys, so patent, we could bleed to death in our fury.

We live with a true, Mephistophelian horror: the belligerent and smug refusal of extractive industries — some food, cattle, and wool growers along with the mining, logging, and hydro industries — to address a deep and terrible damage done to the Earth. Their self-satisfied resistance to change, in which they are enthusiastically, sometimes gleefully, supported by a host of academics and politicians, looms like an epitaph for Western civilization.

Most of us feel powerless in the face of this obduracy. We offer each other as consolation the wisdom and practicality of Aldo Leopold's insights or the morality of Wendell Berry's prose, and wonder endlessly why such thinking finds no welcome in Washington, at stockholders' meetings, or at most academic gatherings. Our trouble, in part, is that we are beseeching institutions intent on their own preservation to become interested instead in preserving something else entirely — life.

It is not likely.

Our answer must be, if we are to find hope in the face of this hopelessness, to turn our backs on the institutions — the plutocrats in politics, the corporate champions of a salted earth, the hyper-righteous lobbyists and other apologists for toxins, clearcutting, nuclear reactors, and overgrazing. A person turns her back by putting a seedling in the earth with her two hands. A person turns his back by firing prairie grass. You turn your back by rediscovering the plants and migratory birds of your own back yard. We turn our backs by joining in the hard, common, physical labor of shoveling, seeding, heaving, portaging, firing, transplanting, and channeling to restore disturbed and abused land to a semblance of its original, complex, graceful coherence.

We turn our backs by accepting the responsibility that government and industry reject. At some point their opposition becomes so willfully ignorant, so wedded to catastrophe, so deranged by a sense of destiny that it becomes irrelevant.

I know of no restorative of heart, body, and soul more effective against hopelessness than the restoration of the Earth. Like childbirth, like the giving and receiving of gifts, like the passion and gesture of the various forms of human love, it is holy. And like David Wingate's Nonsuch Island in Bermuda or the Illinois savanna restored by Steve Packard and his friends or the work of Janet Morrison and her companions on the Mattole River in northern California, the evidence of this act of love is inspiring.

From a practical point of view, restoring watersheds and prairies may be biologically self-serving, but its ultimate value is broader and more profound. At the heart of the phrase ''the Earth teaches'' lies an acknowledgement that human beings alone do not know what needs to be known to lead healthy lives. What needs to be known can only be known in concert with the Earth, by participating in instead of trying to manage our biology.

Restoration work is not fixing beautiful machinery — replacing stolen parts, adding fresh lubricants, cobbling and welding and rewiring. It is accepting an abandoned responsibility. It is a humble and often joyful mending of biological ties, with a hope, clearly recognized, that working from this foundation we might, too, begin to mend human society.

It is no accident that restoration work, with its themes of scientific research, worthy physical labor, and spiritual renewal, suits a Western temperament so well. It offers the promise, moreover, of returning to us what we so persistently admire in indigenous cultures — a full physical, spiritual, and intellectual involvement with the Earth, and an emphasis on the primacy of human relationships over the accumulation of personal wealth.

We can sense, in other words, salvation here. And we can imagine, too, latent in this movement, a potential not simply to change the direction of Western culture but to alter its foundation.

Barry Lopez
McKenzie River, Oregon
February 1991

# INTRODUCTION

Behind the awareness of our environmental dilemmas there lurks a question — how do we change the way we live? Large segments of our economy and our government still dispute the need for change, and view anything different as deprivation. Alter how we live and consume, they say, and we will all be bitching and grumbling our way into the next century. By ignoring the warnings that arrive daily from degraded ecosystems, however, we risk not only our own future, but the survival of much of the environment which surrounds and supports us.

Through the practice of environmental restoration, the people in this book are inventing better solutions. By learning to make adaptive responses, they are helping to improve both the environment and their communities — by first realizing that the environment *is* their community. Nature often works more slowly than the speeds modern humans are used to (quick solutions to biological problems are almost always the wrong solutions), and some of these projects will take lifetimes to complete.

That kind of altruism can be hard to sustain, but fortunately the acts of restoration themselves provide satisfaction. Righting wrongs is inherently ennobling, and in the process, these restorationists are discovering how to live lives that are more connected and rewarding. Their stories display the sense of accomplishment that can come when we align our actions with forces larger than ourselves. I hope their examples serve as an invitation for you to join in and get your own hands dirty.

One of my best childhood memories is of a weekend spent planting trees with the Boy Scouts. The poplar saplings lined a suburban street and grew to be enjoyed by the whole neighborhood, but what most impressed me was realizing how much could be accomplished — and how quickly — by an organized group. It was one of the only times I experienced the excitement of being part of a collective energy aimed at a common task. Group effort like this is a given for a Third World peasant, or for anyone who farmed fifty years ago. We have lost the necessity for such activity, delegating it instead to machines. Not needing to dirty our hands in the earth is taken as a hallmark of our progress, but something vital has been lost. Restoring the environment is a collective physical act that reverses these tendencies. It engenders local responsibility anchored in the future of local places.

Environmental restoration does not replace the need for conservation or preservation. Best to do no harm to the environment in the first place, or to stop harm already under way. But it is no longer sufficient to ignore the mistakes and indifference of the past. Damage has been done, and needs to be repaired. Over half of us in the United States now live in just thirty-nine monster metropolitan areas, and many of the places most in need of restoration are very close to our own homes.

By connecting us to the fate of specific neighborhood ecosystems, restoration work affords the understanding that our own human future abides there as well. A ripple effect can result, and the reasons for making still other changes in our lives become much more tangible. If you are helping to restore fish to a local stream, the whole effort can hang on the acidity of the water. Once you realize this, the pollutants that come from your car's exhaust pipe or the nearest power plant cease to be abstract, and the benefits of using mass transit or buying energy-efficient appliances become easier to see. The synergism of the biological world spills over into our own lives. Nature never operates in isolation, and by working with Her and with each other, we can help to end our own isolation as well.

Richard Nilsen
Editor

# 1.

# THEORY OF ENVIRONMENTAL RESTORATION

ENVIRONMENTAL RESTORATION, like Nature itself, is always changing and growing. The rules of this young profession aren't all agreed upon — even by people who make their living as restorationists. Better techniques for healing damaged landscapes are always evolving. To sketch in the boundaries of this new science, we begin with several different viewpoints, all aimed at giving you a clearer idea of what environmental restoration is, what it isn't, and what it may yet become.

One of the first things to realize is that this is more than just science. Jim Dodge sees it as an art form of the future and ''a new genre of the healing arts.'' William Jordan describes the origins of the profession of restoration on abandoned Wisconsin farmland in the middle of the Great Depression. Seth Zuckerman examines restoration's recent past and, as with any great new idea, identifies its potential for distortion and abuse. Peering beyond the hype and dislocations of modern industrial society, Wendell Berry argues that any kind of environmental work must be local and must also strengthen human communities if it is to succeed in the long run. Susan Davis considers what happens when humans don't intervene, and Nature is left to heal itself. And Michael Ortiz Hill examines his own life in the Nuclear Age, and suggests that restoration can also involve personal healing.

When environmental restoration is most successful, it also improves our hearts, and cultivates an enduring relationship with Nature. That's a hard one for many of us to get, having wandered this far down the technological highway, but it is crucial to our sanity, and vital to this work. Done properly, environmental restoration restores far more than just the land.

# Life Work

*BY JIM DODGE*

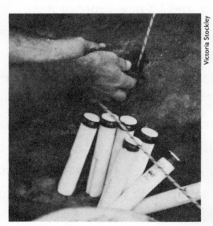

(Above) The Cazadero Forest Workers' crew replanting a burn.

(Right) You can't cut boulders with a chainsaw — setting charge to clear a midstream boulder at the bottom of a log jam.

(Below) Crew member Victoria Stockley finds the missing wedge.

A basic rule of human habitation has always been "Don't shit in camp." When the number of human inhabitants reached five billion, "camp" became the planet, the common ground of the global village and everyone's backyard. This recognition that our species' pollution and pillage in disparate locales now threatens, by cumulative insults, to foul the collective nest has been relatively sudden (the last thirty years of human occupation) and perhaps too late. But if necessity is the mother of invention, species-survival may prove downright inspirational.

The notion of "right livelihood" — work that is individually satisfying and for the common good — is beginning to exert its simple wisdom on a wider scale. If we spend about a third of our lives sleeping and a third at work, a comfortable bed and a fulfilling job are obvious provisions for getting a jump on happiness. A recent Columbia University study of people over 95 years old asked the elders what they'd change if they could re-live their lives. Among the three most frequent answers was "Create something that would last beyond death." (The

The author consults with old-timer Jimmy Kennedy about removing a log jam on Palmer Creek in the Gualala River watershed in Northern California. Some log jams act as barriers to spawning salmon. This one took a crew of eight one week to remove with hand tools.

# Some Reasons Why Environmental Restoration May Be The Art Form Of The Twenty-First Century

Cazadero Forest Workers' crew stabilizing the toe of a cutbank to prevent further erosion. Materials used came from a nearby log jam.

other two were ''Take more risks'' and ''Stop and smell the roses.'') Restoration work not only offers right livelihood at a time when we seem sick-to-puking with meaningless labor for the poverty of dollars, it also, according to our elders, insures us against regret. Work is work, but it's a pleasure to sing for one's supper when the song itself provides sustenance.

Restoration, like any art, seeks a greater understanding of existence, which tends to deepen our appreciation, gratitude, and humility, salubrious states of mind that are less fringe benefits than compelling requisites for further work. Moreover, the art of restoration is finely balanced between mind and body, thought and sweat. The work is heads-up and hands-on: figuring out by guess, gut, experience, and calculation how many trees to plant on the hillside, what species, when, and to what immediate and long-term consequence, then picking up your hoe-dad and doing it. Even better, it's outside work, so you don't have to make a special trip to smell the roses or feel the rain on your face. Compared to restoration, the other

arts seem dangerously self-involved and boringly one-dimensional.

It's my feeling, one that swings between conviction and wild hope, that we're at a point in species' consciousness where we're about to grasp both the importance and rightness of helping the planet heal the injuries we've mindlessly inflicted. It's tempting to think of restoration as a new genre of the healing arts, but in fact we have hardly begun to wash the wounds, much less address the psycho-social pathologies and self-destructive contempt that fuel the affliction. In caring for the earth, we may heal ourselves, but at present the art of restoration is more janitorial than medicinal. We've made a mess of creation, and now we need to clean it up. ■

*Jim Dodge is the author of* Fup, Not Fade Away, *and* Stone Junction. *He is a former restoration worker and a running-dog bioregionalist who lives in Arcata, California.* —Richard Nilsen

University of Wisconsin

# Restoration:
## Shaping The Land,
## Transforming The Human Spirit

## BY WILLIAM R. JORDAN III

*Bill Jordan is a staff member of the University of Wisconsin Arboretum, where he is editor of* Restoration & Management Notes. *He is also a founding member of the Society for Ecological Restoration (see p. 76 for reviews).* —Richard Nilsen

**(Above) The Curtis Prairie, the oldest example of restoration in the United States, was reclaimed from farmland in 1934.**

**(Below) Labor was provided in the early days by the Civilian Conservation Corps. In the summer of 1936, buckets were used to keep young plantings alive.**

University of Wisconsin

Ecological restoration has only recently begun to attract the attention of large numbers of environmentalists concerned about the conservation of natural areas, but neither the idea nor the practice of restoration are entirely new. What is now generally regarded as the first systematic attempt to restore native ecological communities on disturbed land began in 1934, when a faculty committee at the University of Wisconsin in Madison proposed developing a collection of native ecological communities on derelict farmland then being acquired by the university for an arboretum.

The purpose of this project, as expressed by Aldo Leopold, who played a leading role in developing and carrying it out, was to establish a collection of all the ecological communities native to the area. To some extent this could be done simply by protecting existing communities from further disturbances. But since many of the original communities no longer existed on the site, it would be necessary to recreate them. The attempt to do this, carried out during the first half-dozen years by Civilian Conservation Corps enrollees working out of a camp on the site, and in more recent years by a succession of faculty and students, has gradually led to the creation of the most extensive collection of restored ecological communities in the world — a total of over thirty more-or-less distinct community types covering roughly 300 acres and representing all the major communities native to Wisconsin and the upper Midwest.

For many years the Arboretum served as a model and training ground for the few conservationists involved in restoration work. More recently it has played a leading role in the development of restoration as a discipline. Broadly speaking, there have been two phases to the development of ecological restoration.

In the first phase, the emphasis was on the product of the restoration effort — the restored community as an object in the landscape. This was only natural since it is precisely a concern about the quality of the product (specifically, its authenticity, or resemblance to the natural or historic system chosen as a model) that distinguishes restoration from other forms of land rehabilitation. Yet taken by itself this conception of restoration is limited in its value as a strategy for natural area conservation. This point of view simply ignores what may be restoration's greatest benefit — its value *to the restorationist* as a process, a way of learning about the system being restored, and finally a way of establishing an intimate, mutually beneficial relationship with it.

This more recent conception of restoration began during the late 1970s as a direct result of our need to explain the significance of the Arboretum to the public and also to find ways of maximizing its value as a facility for teaching and research. In both areas, the partial artificiality of the restored communities was something of an embarrassment. Fifty years earlier, Aldo Leopold and his

University of Wisconsin

This photo from the 1940s shows a controlled burn, necessary for the prairie's survival. The man on the left is Aldo Leopold.

colleagues had proposed creating a collection of communities which ecologists could then study in place of their natural counterparts, which were becoming increasingly rare. The problem was that even after a half-century, none of the restored communities was fully authentic, and this clearly limited their value for both research and education.

One midsummer day during my first year at the Arboretum I was walking on Curtis Prairie, reflecting on its deficiencies and defects, when it occurred to me that from a certain point of view those defects were actually the most interesting things about the prairie. If there were still defects in

these plant communities despite decades of our tinkering, perhaps they were telling us something about our limitations in understanding them. Perhaps what was really interesting here was not the communities themselves as more-or-less finished products, but the process of restoring them. Maybe the best way to learn about a community like the prairie was to try your hand at restoring one.

Over the years some of the most interesting research from the Arboretum had been a direct result of the restoration work, and would have been difficult or impossible to carry out in an entirely "natural" landscape. An obvious example was the "discovery" of the prairie fire back in the 1940s. Though the prevalence of fire in tallgrass prairies in historic times was well documented, its role in the ecology of prairies was not clearly understood. Restoration efforts at the Arboretum began without employing fire, and the results were highly unsatisfactory. Reintroduced prairie natives flourished, but so did a host of weedy, non-prairie species, which prevented the natives from closing together to create a real prairie. Research carried out on the Arboretum prairies soon established the critical role of fire in the ecology of the tallgrass prairies of the upper Midwest.

Along with a number of similar experiences at the Arboretum over the years, this suggested a pattern in ecological research in which ideas based on observation and analysis are put to the critical test of synthesis. This soon led to the notion of restoration as a technique for basic research, a way of raising questions and testing ideas about the systems being restored. We even coined a term — "restoration ecology" — to refer to this approach to ecological research, and in 1984 the Arboretum sponsored a special symposium to explore this idea as part of the celebration of its 50th anniversary.

Here was a first step toward serious consideration of the process of restoration and its implications for the restorationist. The notion of restoration ecology suggested at the very least that restoration was a powerful

way of learning about a natural system, establishing, if you will, intellectual intimacy with it. The next step was to ask just what kind of activity restoration is, and to recognize its affinity with agriculture, medicine and art. Looking back, I realize that what we were doing was learning about the peculiar relationship of the restorationist to the landscape. We were looking at the process of restoration as a ritual and "reading" it to see not just what it accomplished or produced, but what it expressed about a person and his or her relationship to the land.

Watching a group of volunteers collecting seed on Curtis Prairie one fall day, I realized that they were repeating the experience of hunter-gatherers who inhabited this area centuries ago, and who actually, through their hunting, gathering and burning, had helped create the prairie communities we tended to think of as "native," "original," or "natural." At this point I realized that restoration represents a reenactment — not only of the forces that created the communities being restored in the first place, but of the entire passage of cultural evolution, from hunting and gathering through agriculture, to the analysis and synthesis of modern science. I now see restoration as providing the framework for a system of rituals by which a person in any phase of cultural evolution can achieve a harmonious relationship with a particular natural landscape.

Looked at in this way, it is clear that the business of ecological restoration is much more than a technical and scientific challenge. It is also a model for a healthy relationship between ourselves and nature and, beyond that, a way of exploring, defining, and ultimately celebrating the terms of that relationship.

It is here that I now see the greatest value of restoration — not in its ability to transform the landscape directly (or at least not only in that), but in its ability to transform it indirectly through the education and transformation of the human beings who inhabit and shape it. ■

# PITFALLS ON THE WAY TO LASTING RESTORATION

## BY SETH ZUCKERMAN

COLOGICAL restoration jumped out of the wetlands and prairies and onto the national agenda in 1988, when the Restoring the Earth conference drew an overflow crowd of 800 scientists, policymakers, and activists. A professional society was formed and "restoration ecology" even gained its own entry in *The Reader's Guide to Periodical Literature*. Since the Berkeley conference, practitioners of restoration have grown increasingly introspective about the proper aims, methods and values with which to pursue the discipline. This article explores some of the pitfalls on the road to lasting restoration.

*With two scalping strokes, my hoe slices through a patch of thick sod high on Bear Trap Ridge. It is mid-February, and wind stirs the drizzle around this steep grassy hillside on California's North Coast. I work among a crew of half a dozen, reforesting land that ranchers long ago cleared of trees to create pasture for their cattle and sheep. I shave squares of grass at their roots so the seedlings we plant will have a headstart on their competition. As I move a few steps forward to the next planting spot, I recall the friends who have scoffed that such backyard restoration projects are of little significance next to global problems such as the fraying ozone layer, the threat of nuclear war, or the greenhouse effect.*

This criticism of ecological restoration is common among

*Like any good idea just catching on, environmental restoration can be misused. Progress to one group may look like co-optation to another. This article explores the motivation behind restoration work and reveals that both the science and the politics involved require thoughtful players.*

*Seth Zuckerman is an environmental writer living in Petrolia, California.* —Richard Nilsen

people involved in traditional environmental or leftist organizations whose scope is the entire country, continent or planet. As they say, it would take hundreds of billions of trees over the coming decades to suck carbon dioxide out of the atmosphere and slow down global warming. But even so ambitious a project would be made up of smaller parts. One can estimate the number of trees needed to absorb the carbon dioxide created by a lifetime of fossil fuel use. For an average resident of the United States, that turns out to be about 1,500 trees, if they grow to be a hundred feet tall — as many as I planted in three days or my speedier colleagues in two.

Perhaps tree-planting is too easy an example of local actions affecting global problems. Consider instead genetic diversity, diminishing rapidly as habitat vanishes. In our watershed, we trap salmon and incubate their eggs to help a native strain of fish survive. This population of salmon has adapted to local conditions and evolved within our watershed separately from its neighbors. Of such little pieces is the preservation of genetic diversity made. Our watershed, 92 percent of whose timber has been logged in the last forty years, is far from pristine. My neighbors and I face the task of restoring what has been trashed. Legislation — such as designating a Wild and Scenic River — may protect larger habitats in one swoop than I can in my figurative backyard. But living where I do, *these* actions are available to me.

We each have our niche: I'm glad that people are doing battle in the legislative labyrinths, strategizing continentally and thinking globally — as long as they don't tell me not to act locally. Indeed, what alternatives do I have? Working full-time for a lobbying or policy-oriented group would mean leaving the place where I live. Even if I lived elsewhere, not everyone can enlist in the brigade of full-time environmental activists. Another option — simply donating money to support these good works — rings hollow to me. If that were all I did, I would feel one step removed from the land, lacking visceral connection to it, as if I were in a long-distance relationship with the place I care about. Eco-lobbyists and tree-planters each work for the planet in their own ways. We must collectively grasp the computer in one hand and the hoedad in the other.

*A dozen more planting sites laid bare now, I look back and see that my scalping partner and I have put some distance between us and the planter who follows behind with the seedlings. As I near the crest of the ridge, I can see across the valley to the next slope, thickly wooded with Douglas fir and tanoak, as this ridge will be in a decade or two. It pleases me to think that this local project fits into a bigger picture of healing the earth's atmosphere. But then I remember a patch of old-growth forest near the headwaters of the watershed, which is facing the threat of immediate logging. Is it a waste of time to be out here planting two-year-old Douglas fir saplings for the Bureau of Land Management, instead of helping out with the lawsuit against the timber company to preserve centuries-old trees?*

This is a familiar argument against restoration: to hold

off putting our energies into it until everything possible is preserved. Advocates of this position, including such prominent activists as Sierra Club Executive Director Michael Fischer, argue correctly that an ecosystem preserved intact has greater value than a place that has been restored. Why, then, practice restoration now? Why not give everything we have to the preservation of remaining wild ecosystems?

One answer is that some restoration can stave off even worse damage. Much of the rehabilitation work being done in logged-over portions of Redwood National Park, for instance, will keep soil on hillsides to grow trees, grass and elk, instead of washing into streams and mucking up the salmon's spawning gravels (see page 59). It would take centuries to build that soil back again. Other restoration work — like David Wingate's work on Nonsuch Island in Bermuda (see page 78) — expands the habitat of endangered species, thereby preventing extinctions that might otherwise occur. And restorers typically agree that restoration must be pursued side-by-side with preservation.

Another answer lies in the emotional effect of the two activities. Restoration gives me a feeling of constructive connection with my home place that no amount of lobbying can produce. After a day of tree-planting, I feel physically exhausted but spiritually invigorated. After a day of writing comments on Environmental Impact Reports, I feel drained and useless, in need of a good hike. The level of desperation in the two activities is different, too. If we lose one preservation battle, we have usually lost for good. But in restoration, we often have the flexibility to try diverse ways to reduce erosion from a bare slope, for instance, until we find a way that works. Restoration, says Peter Berg of Planet Drum Foundation, is like adding a

birthing room to a hospital that had only a trauma center.

Activists and inhabitants alike have taken up this calling for the long haul. Like cross-country runners, we must pace ourselves. Unless we care sustainably for the places where we live, we'll burn out long before the work is done, and without having encouraged the next generation to carry it on. Besides, there's a lot to be learned in the pioneer stages of restoration work. In 2010, when Fischer says the preservation battles will be over and restoration should begin, inhabitants of places will have a better base of knowledge and experience from which to restore if some of us have already gotten our hands dirty.

*High on the ridge, the soil dwindles and my hoe scrapes gravel with nearly every stroke, no matter how shallow. Not a great place to plant a tree, nor a great place to grow if you are one. The wind blows fiercer up here, leaving the view as the only advantage, a skyline of hilltops and ridgelines. I feel lucky to be able to forge this kind of link with the land, to take part in returning this land to forest. Some of my friends who live in the city envy me this opportunity, thinking that nothing like this is available to them. I keep telling them it's a matter of finding nature wherever it may be, among the coffeehouses, bookstores and housing projects.*

Restoration is not the prerogative of rural areas alone, although it may seem a daunting task to metropolitan dwellers when they see how little of their native ecosystem remains among the asphalt, tract homes, and shopping malls. Native prairies have been reestablished in Chicago and Lawrence, Kansas. Because the native grasslands of the Great Plains coevolved with fire, restorationists burn the prairies there every year or two, inside the city limits. Another indigenous grassland is being painstakingly restored by volunteers on the windswept top of Bernal Heights in San Francisco. Two small Philadelphia firms,

Natural areas are where you find them. San Francisco's Bernal Heights is a grass-covered rock hill topped by microwave transmitters. Volunteers are reintroducing perennial native bunch grass species which were displaced when the Spaniards brought annual European rye grass 200 years ago.

Barbara Pitschel

*Lasting restoration does not attempt to restore an ecosystem to a particular state; it attempts to restore the processes of natural succession and evolution that occur in wild, self-regulating systems.*

Sere Ltd. and Andropogon Associates, restore native woodlands, meadows and stream margins in urban and suburban environments. Stacy Levy of Sere describes a lot of her work as "editing out" the exotic tree species to leave the natives intact and make the habitat more accessible and less forbidding to people. She has designed projects such as these for the edges of a main thoroughfare leading into Philadelphia and for Dupont's headquarters in Wilmington, Delaware.

Urban creek projects — alive in cities such as Vancouver, B.C., suburban Seattle (see page 123), and Berkeley — seek to bring creeks out of storm sewers or concrete channels, eventually to flow between banks overhung with native alders, willows and ferns. The restoration of the Nashua River in industrial New England has been chronicled by John Berger in his book *Restoring the Earth*.

Not only is urban restoration possible, it holds special educational promise. Most people on this continent live in cities. As long as natural ecosystems are excluded from cities, urban dwellers' only experience of nature will be as spectators in zoos and arboreta or on trips to the country, where they are visitors in an ecosystem far from home. This distanced relationship is unlikely to lead human inhabitants to conceive of Nature as being real, or to care about her as part of themselves; it will not lead them to membership in the land where they live. Daily contact could.

*The planters have run out of trees and we hike back to the truck to get more. Sandwiches and thermos bottles appear, and we recharge. After the break, it's my turn to plant, and I take bundles of foot-tall seedlings from the bag and dip their bare roots in water to keep the delicate rootlets alive. Then up the hill, to get 'em in the ground and make those trees grow. Wait a minute. Make those trees grow? I kick myself for the arrogance in that thought. Who do I think I am, Gaia herself?*

This was not the first time I had been complicit in this pitfall, the treatment of restoration as an exercise of mastery over nature. The first restoration article I worked on, in a 1988 *Newsweek*, proclaimed, "the fix-it men of the environment are here," as if an ecosystem were a washing machine or an automobile transmission. The very word "restoration" carries its share of misleading baggage: the connotation that we can return an ecosystem to a static state it occupied before human disturbance. Further, in the usage, "we restore the stream," we act upon the stream, and the stream simply lies back and is acted upon. This implied relationship attributes omnipotence to us and excludes the human actors from also being transformed and acted upon. It manifests itself in such locutions as the title of the September 1989 special issue of *Scientific American:* "Managing Planet Earth," which imputes to humans the power to manage the planet — as if the workings of Nature lacked mystery. This attitude peaked in Walter Truett Anderson's 1987 book, *To Govern Evolution* (see p. 26). Try these sentiments: "We have made the world over once and it is time to make it over again." "I am not here to argue that the human

A new gas line crossed a park in Morris County, New Jersey. Andropogon Associates modified a tractor's scoop bucket so it could remove 4-by-8-foot mats of native vegetation before the trench was dug. The center photo shows a mat being slid off its pallet and into its new home on top of the buried pipe. The left photo shows site work on a small stream during construction, and on the far right, the results six months later.

Andropogon Associates

species ought to take responsibility for evolution on the planet. . . . That is not the question before us since we are already governing evolution.''

We have little reason to be so confident. Previous attempts to control the Earth have disrupted natural systems, driven species to extinction at a rate unprecedented since the dinosaurs died out, and done numberless injustices to countless human cultures. Stewart Brand once wrote, ''We are as gods and might as well get good at it.'' I prefer Anne Herbert's restatement in *The Next Whole Earth Catalog*, ''We are as humans and might as well get used to it.'' The record of attempted restorations in the next section suggests that restoring the Earth will not be as straightforward as restoring a Model T. Let us learn to work with Gaia, not apart from her. Hominids are but a blip on the evolutionary landscape; farmland or logged areas are potholes in the surface of ecological succession, albeit sometimes gaping ones. One challenge before restorationists is to blend our efforts with these far greater forces, and to become a part of them, and not, in the words of the Book of Genesis, to ''have dominion over the fish of the sea, and over the birds of the air, and over every living thing that moves on the earth.''

Hiking along a logged-over creek in California's Mattole River valley, a student of mine and I pondered the recovery that had occurred since the forest was cut twenty years before. Gradually, trees were growing back, shade was

returning to the streambanks, and the creek was cleaning out the silt and gravel that had eroded into the channel. Scars still showed: bare soil, a road slipping into the creek, a forest dominated by the hardwoods the loggers had ignored, instead of by Douglas fir. But it was clear that, left to its own devices, the watershed would recover. Fire would beat back the tan-oak and madrone, stray salmon would recolonize the stream (assuming that some fish-bearing streams remained on the North Coast), natural succession would resume where it left off. Perhaps on its own it would take a thousand years; perhaps with human help it would only take four hundred.

Does the watershed care how long it takes? Probably not much. It might help preserve a native strain of some species, or pull carbon dioxide out of the air sooner if we actively restore, but the watershed will eventually return to much the same successional processes if humans walk away. So why should human inhabitants care? Not so much because we want it to look better in our lifetimes or those of our grandchildren, but because the very process of restoration heals our relationship to the place where we live. Try it. Plant a tree on your street. Hear the rhythms of your life-place, imagine what would harmonize with them, and do it.

Several of my friends volunteer at Martin de Porres soup kitchen in San Francisco. Every so often, a new volunteer arrives who believes that he's helping the hungry out of some sort of noblesse oblige. It doesn't take long before he realizes that he's not there to help them, he's there to help himself. So it is with lasting restoration as well.

*Thwunk! My hoedad sinks a few inches into a patch of newly bared earth, but not deep enough for the roots of the tree. I straighten up for another swing of my narrow-bladed hoe.* ►

The once and future meadow.

Years ago, to mine the minerals underneath this meadow in eastern Utah, someone took away the topsoil and didn't put it back.

And the life that depended on it walked a surface as barren as the moon.

Then, the land changed hands. And the people who bought it did what should be done every time you mine the land.

They put back topsoil, plowed furrows, planted grass.

And brought back this meadow where life once flourished and now does again.

Do people really repair the past just so nature can have a future?

People Do.

Chevron

Do people lie to you on television about their corporate activities, so you won't object when they come to wreck your neighborhood?

People do.

Do people write advertisements about endangered species protection for multi-national oil companies, who are among the worst polluters on the planet?

Chevron

PEOPLE DO.

*Thwunk! This time, it goes in more than a foot, up to its hilt. At least I believe in this, I think. I'm not doing it merely to create a monoculture of Douglas fir, a timber plantation of identical trees marching from one ridge to the next. The trees that used to stand here weren't cleared with the excuse that they could always be replanted — they were cut and cleared by ranchers who saw their livelihoods in it, at a time when American society wasn't thinking about the reasons not to.*

One of the deepest, slipperiest pitfalls of restoration is an attitude that every ecosystem can be replaced. Restoration could thus serve as an excuse to destroy inconvenient habitats — such as wetlands located precisely where a developer wants to put an industrial park — because new habitats could be created elsewhere. Such habitats can satisfy the demands in the National Environmental Policy Act and comparable state laws for "mitigation" of environmental insults.

What's wrong with thinking one can perform excellent mitigation? For one thing, it can slop over into the attitude of mastery over nature described before. Ecosystems are precious assemblies of life-forms, each uniquely adapted to its specific site. People fail to acknowledge those characteristics if they treat a habitat like a house that can be jacked up and moved to a new location. It is as if a doctor suggested breaking someone's arm because, after all, he knows how to set broken bones. "The purpose of restoration is to repair previous damage," cautions author Berger, "not to legitimize further destruction."

You needn't accept this philosophical argument to deplore the use of restoration simply for mitigation's sake. We are not as gods. We cannot infallibly create new habitat, nor can we decree that the birds, fish, and invertebrates will inhabit what we create. After all, our species has trouble building public housing that anyone wants to live in. Who will presume to understand other species better than our own?

Examples abound of misguided, over-trumpeted restora-

Ask a rhetorical question in an advertising campaign (top), and you are bound to elicit unsolicited responses (middle and bottom). Chevron's "People Do" TV and print campaign highlights environmental good works by a caring giant oil company. The middle reaction is from a Northern California group opposing offshore oil drilling; the parody below appeared in *Processed World*, a satirical magazine for post-industrial wage slaves.

The question remains: if you always assume the worst about corporate public relations, how do you encourage big business to do a better job caring for the environment?

# If you do a truly good job of restoring an ecosystem, it will decide for itself what it grows into after you've gone.

tion projects. In North Carolina, several years after the apparently successful restoration of a wetland, only a sixth as much living matter was produced each year as in a comparable natural marsh. More disturbing, the species composition was quite different — fewer mussels and marine worms, many more insect larvae and amphipods. So although the marsh seemed to be restored, it was ecologically quite different from a natural one. In San Francisco Bay, restoration projects have flooded non-tidal salt marshes, destroying one kind of scarce habitat (badly needed by the endangered salt marsh harvest mouse) to create mediocre facsimiles of tidal salt marshes. Dr. Margaret S. Race, writing in the journal *Environmental Management*, cites numerous instances of reports exaggerating the success of restoration projects. In one case, a marsh just beginning to be colonized by vegetation and needing major channel work was characterized as a successfully restored waterfront park. Another wetland was said to "demonstrate that a marsh can be established in two years," when aerial photos showed an irregular patchwork of planted plots (covering less than a tenth of the site) and open mud.

Nor is this inflation of results solely the province of wetlands restoration. Mineral companies, now required by federal law to replace topsoil after they strip-mine and replant the land, boast how much better they perform than in land-raping days of yore. Chevron's "People Do" advertising campaign touts its mine-reclamation efforts on land in Utah that it is not legally required to reclaim. But a shortcoming in Chevron's work and in much strip-mined land reclamation is the type of vegetation the mining companies re-establish. They plant grasses, which are most likely to provide the quick ground cover the law requires, and which do slow erosion from the re-assembled hillsides. But it will be seventy years before the native pinon and juniper return to Chevron's Utah site, according to the company's environmental security specialist John Laursen. Back east, native trees such as pin cherry and aspen recolonize some unreclaimed mined land within a few years, says Assoc. Prof. H. Glenn Hughes of Pennsylvania State University. But after seeding, grasses often create tight sod that prevents native trees from reseeding the site. After eight years, Hughes says, nearby trees still have not encroached onto the supposedly restored areas, leaving an ecosystem that is quite different from what was there before.

With all of these foibles in restoration work, its use as mitigation must be approached with great caution. Planted pickleweed doth not a pristine marsh make. Accepting a restored ecosystem in partial trade for the destruction of another ecosystem means swallowing a decline in productivity, a radical change in species composition, or the possibility of outright failure. Nonetheless, lasting restoration has a place in balancing activities that involve habitat destruction. Homestake Mine has undertaken a project at its McLaughlin gold mine near Clear Lake, California, that stands as an example to aspiring restorative miners. Homestake is responsible for the management of 10,000 acres, of which only 1,200 are actually used in mining operations. The land was mined for mercury for over a hundred years, beginning in the 1860s. Vast tracts of the native oak woodlands were cut down, and the streams were fouled with mercury and toxic waste from the extraction process. The land around the mines was overgrazed and came to be covered in tarweed and star thistle, plants that the cattle don't eat.

When Homestake came on the scene in the early 1980s, it reduced the number of cattle on the land to a sixth of their former number. It minimized the mine's impact by such means as siting the ore processing five miles from the mine pit to locate the tailings pond over impermeable rock, where its contents wouldn't leach into the groundwater. Even though the mine moves fifty thousand tons of rock a day, air-pollution levels are lower than over most agricultural land, because of stringent dust-control measures. Skeptics point out, rightly, that Homestake might have been denied its permits if it didn't make such an extraordinary effort, and that the reclamation efforts on mined land are required by law. But Homestake is doing more than just planting rye grass on the areas covered by topsoil and waste rock. It hired a local restoration firm to collect native seed from brush and tree species, and propagate them on the reclaimed areas. Besides, Homestake can justly take credit for restoration work above and beyond what it originally committed to doing, and for an excellent plan to restore habitat on the land it manages that is not directly affected by its mine. It bought some abandoned mercury diggings upstream of its reservoir that were leaking the metal into the creek below, and is diverting runoff around it, burying the tailings and revegetating them. "The staff is proud of the environmental reputation this mine has," says Ray Krauss, the mine's environmental manager.

So far, most restoration work is fairly low-key. "The philosophy of our plan is to halt all human intervention and monitor the consequences of natural succession," explains Krauss. "Then we see to what extent natural processes don't remedy historic problems, and only then do we intervene." Data from the last few years have convinced Krauss to plant riparian vegetation along the margins of the reservoir; elsewhere, the land already appears to have

Homestake Mining Company's McLaughlin gold mine in Northern California moves 50,000 tons of rock each day. Much of their restoration work involves stabilizing and revegetating these tailings. Going beyond minimum regulations, Homestake grades to proper slope (right), then seeds with grass (middle), and finally plants native shrubs and trees propagated from seed gathered locally (far right). When the gold is gone in twenty years, this will all become an environmental research station.

Seth Zuckerman

tremendous value for wildlife: golden eagle, bear, bobcat, and mountain lion are seen on the unmined land, which adjoins a large BLM wilderness study area. Aside from the exemplary work that Homestake is doing in reclaiming the land it has disturbed, it shows how some $2 million of the mine's $80 million annual revenue can finance the rededication of a vast tract of land to the wild. A final note separating Homestake's efforts from mere mitigation is the intent and attitude behind it: Homestake is not grudging in its support for the local ecosystem, nor does it seem to be doing this in a short-term attempt to manipulate the political process. Instead, it constructively and generously engages the local biota. At the end of the mine's twenty-year lifetime, when the site is dedicated as an environmental research station (complete with two decades of baseline data), it will be in much better shape than when Homestake arrived.

*A few trees later, my hoedad is once again buried up to its hilt. I pull up on the handle, widening an underground opening for the roots of the seedling and making sure the soil is well-loosened. I reach into the bag around my hips and pull out a tree, green needles above and damp bare brown roots below. I pull back on the hoedad, opening a narrow slit in the earth, and twist the seedling as I lower it into the opening so its roots snake straight down the hole. In a decade, I'll be able to see these trees from the porch of the general store. Tree by tree, we who dwell here are making this our valley, and the valley is making us its humans. On a less philosophical plane, we locals are getting the 27 cents for each tree we plant.*

With the rise of restoration as a media topic, talk has circulated of a national restoration project, much like the Depression-era Civilian Conservation Corps. It's fine to push for state or federal restoration funding, but we don't need enlightened technocrats to strategize the recovery of eroded farmland and depleted salmon streams, all the while pitying the so-called yahoos whose ignorance caused these problems in the first place. Such an attitude won't lead to lasting restoration, and will sabotage the interests

of all people in restoring the biosphere. Instead, let the work be directed as much as possible by the inhabitants of the places to be restored.

One example of the importance of local initiative and participation comes from the experience of urban tree planters. Brian Fewer, former head of urban forestry in San Francisco, recalls that only half of the street trees planted under a federal summer-jobs program survived. Later, when the residents of the neighborhoods took part in the planting, survival topped 90 percent. This parable doesn't mean there is no place in lasting restoration for outside help. Freeman House (see page 46) describes the appropriate assistance the Mattole salmon group has received from the California Conservation Corps and Redwood National Park.

A difficulty arises in deciding which ecosystems shall be restored. In the Summer 1989 issue of *Earth Island Journal*, Gar Smith urges an approach of "environmental triage" — choosing the most important ecosystems and focusing on preserving or restoring them. Such prioritization can help allocate scarce resources more effectively among different possible projects, but it immediately raises the question of who is to decide. It's hard to imagine Smith's approach without a centralized decision-making body that, history teaches us, is unlikely to include grassroots representation.

This notion of triage fails to take full account of a hidden, nearly untapped reservoir of energy that can be brought to bear for restoration by the inhabitants of places who are likely to feel an attachment to their homes that they may not feel to the rainforest or the river delta. The state bureaucracies had written off the Mattole salmon runs as too far gone to save. But the people who live here decided otherwise, and have devoted incredible energy to the fish, so far staving off the run's demise.

Smith argues that "it makes no sense to repaint the kitchen cabinets when the house is on fire." Undoubtedly,

Seth Zuckerman

Seth Zuckerman

major threats to ecosystems — oil spills, climate change, chemical disasters — need to be addressed while restoration efforts proceed. But some people's role in the fire brigade will be to restore places near their homes, helping Nature reassemble the biological integrity of their region.

*I pull out the blade of the hoedad, and crumbs of soil slide down into the hole. With the hoedad and then my boot, I tamp the earth around the tree to make sure no air pockets are trapped near the roots. I stand back for a moment to admire my handiwork. To my right, a faster planter has worked ahead, dotting the hillside with seedlings in a neat honeycomb pattern, eight feet between trees. The short green trees in their splotches of bared brown soil look somehow pleasing to my eye. But what can aesthetics tell me?*

If we adopt the goal of restoring the ecological integrity of a place, aesthetics cease to be useful as soon as aesthetic judgments cease to coincide with ecological ones. No matter how many people believe that vast expanses of year-round green lawn are beautiful, it still isn't restoration to establish them in California. Lawns are not part of California's natural ecosystem, no matter how much they tickle American sensibilities. A fifty-year-old Douglas fir plantation may look attractive to an untrained eye, but if it replaced a mixed-age, mixed-species forest, it is a poor excuse for restoration. Rather than throwing out aesthetics altogether, let us adjust our aesthetic to recognize the beauty in the landscape that Gaia creates.

That landscape changes continually. A fundamental problem with aesthetic criteria for restoration is that they can lead people to treat ecosystems as canvasses for their artistic visions. People can thus seek to restore the land to a particular snapshot of ecological beauty from an imagined earlier era: before industrial logging, before the arrival of white settlers, before the diking of the wetlands. But any such snapshot only existed for a moment in time. Lasting restoration does not attempt to restore an eco-

system to a particular state; it attempts to restore the processes of natural succession and evolution that occur in wild, self-regulating systems. Consider a meadow that occupied a certain site in the 19th century when white settlers first described a place. Fifty years before, it might have been a beetle-infested stand of pine; it might well have burned a decade later. Left to natural succession, and depending on the pattern of fire, it might have become a brushfield and then a forest within a few more decades. In a moister climate, the meadow might have been on its way from lake and bog to forest.

To come on the scene and restore and maintain it as meadow defies the inherently changing order of nature. (Exceptions may be appropriate where rare habitats and the native species they support are in danger of disappearing, if the biome is so reduced or disturbed that new islands of habitat won't appear.) Attempting to preserve an ecosystem at a particular state is akin to trying to hold back a flood with a raincoat. This is a major difference between restoring an ecosystem and restoring a Victorian house: You know how you want the house to look, and what shape you want it to stay in. But if you do a truly good job of restoring an ecosystem, it will decide for itself what it grows into after you've gone.

*Even though my motivations transcend the merely aesthetic, something bothers me about what we are doing on this hillside. Is it that we are accelerating succession and skipping the brush stages that often precede forest? No, the forest soil is still intact despite the clearing. Gazing at the next ridge over, I notice the species of trees that throng together there — madrone, tan-oak, canyon live oak. Here we are, planting nothing but Douglas fir — almost as if we are timber companies with nothing on our minds but future harvest. A monoculture. Hmm. In all likelihood, I rationalize, not all of our trees will live, and oak and madrone will eventually sneak into the gaps between the Douglas-fir. I head on to plant the next tree. ■*

Secrets of the Old Growth Forest

# WORD AND FLESH

*BY WENDELL BERRY*

More than one person has written to warn us at our magazine, Whole Earth Review, *that our human foolishness is creating environmental problems that will harm Gaia. James Lovelock and Lynn Margulis's Gaia Hypothesis is named for the Greek goddess of Earth, and suggests that all life acts as a self-regulating system by controlling our planet's atmosphere. Since our magazine helped present this idea to a general audience back in 1975, this seems a good time to point out that these letter writers have it wrong. The messes we are making may well harm humans (they already are), but they will not harm Gaia. If it comes to it, Gaia can get along without humans just fine, thank you.*

*This also seems like a good place for a few words from Wendell Berry, to help give the notion of environmental restoration a context and a proper sense of scale. This commencement address to graduates of the College of the Atlantic in Maine is excerpted from Berry's collection of essays,* What Are People For?, © 1990 *by Wendell Berry. Published by North Point Press and reprinted with permission.*

*—Richard Nilsen*

**I**T IS CONVENTIONAL AT GRADUATION EXERCISES to congratulate the graduates. Though I am honored beyond expression by your invitation to speak to you today, and though my good wishes for your future could not be more fervent, I think I will refrain from congratulations. This, after all, is your commencement, and a beginning is the wrong time for congratulations. Also I know enough by now of the performance of my own generation that I look at your generation with some skepticism and some anxiety. I hope that in fifty years, having looked back at the lives that you are now commencing, your children and grandchildren will congratulate you.

What I want to attempt today is to say something useful about the problems and the opportunities that lie ahead of your generation and mine. I know how desirable it is that I should speak briefly, and I intend to do so.

Toward the end of *As You Like It*, Orlando says: ''I can live no longer by thinking.'' He is ready to marry Rosalind. It is time for incarnation. Having thought too much, he is at one of the limits of human experience, or human sanity.

If his love does put on flesh, we know, he must sooner or later arrive at the opposite limit, at which he will say, ''I can live no longer without thinking.''

Thought — even consciousness — seems to live between these limits: the abstract and the particular, the word and the flesh.

All public movements of thought quickly produce a language that works as a code, useless to the extent that it is abstract. It is readily evident, for example, that you can't conduct a relationship with another person in terms of the rhetoric of the civil rights movement or the women's movement — as useful as those rhetorics may initially have been to personal relationships.

The same is true of the environment movement. The favorite adjective of this movement now seems to be ''planetary.'' This word is used, properly enough, to refer to the interdependence of places, and to the recognition, which is desirable and growing, that no place on the earth can be completely healthy until all places are.

But the word ''planetary'' also refers to an abstract anxiety or an abstract passion that is desperate and useless exactly to the extent that it is abstract. How, after all, can anybody — any particular body — do anything to heal a planet?

Nobody can do anything to heal a planet. The suggestion that anybody could do so is preposterous. The heroes of abstraction keep galloping in on their white horses to save the planet — and they keep falling off in front of the grandstand.

What we need, obviously, is a more intelligent — which is to say, a more accurate

— description of the problem. The description of a problem as "planetary" arouses a motivation for which, of necessity, there is no employment. The adjective "planetary" describes a problem in such a way that it cannot be solved.

In fact, though we now have serious problems nearly everywhere on the planet, we have no problem that can accurately be described as "planetary." And, short of the total annihilation of the human race, there is no planetary solution.

There are also no national, state, or county problems, and no national, state, or county solutions.

That will-o'-the-wisp of the large-scale solution to the large-scale problem, so dear to governments and universities and corporations, serves mostly to distract people from the small, private problems that they may in fact have the power to solve.

The problems, if we describe them accurately, are all private and small. Or they are so initially.

The problems are our lives. In the "developed" countries, at least, the large problems occur because all of us are living either partly wrong or almost entirely wrong. It was not just the greed of corporate shareholders and the hubris of corporate executives that put the fate of Prince William Sound into one ship; it was also our demand that energy should be cheap and plentiful.

Our economies of community and household are wrong. The answers to the human problems of ecology are to be found in economy. The answers to the problems of economy are to be found in culture and in character.

To fail to see this is to go on dividing the world falsely between guilty producers and innocent consumers.

The "planetary" versions — the heroic versions — of our problems have attracted great intelligence. But these problems, as they are caused and suffered in our lives, our households and our communities, have attracted very little intelligence.

There are some notable exceptions. A few people have learned to do a few things better. But it is discouraging to reflect that, though we have been talking about most of our problems for decades, we are still mainly talking about them. We have failed to produce the necessary *examples* of better ways. The civil rights movement has not given us better communities. The women's movement has not given us better marriages or better households. The environment movement has not changed our parasitic relationship to nature.

The reason, apparently, is that a change of principles or of talk or of thought is impotent, on its own, to change life.

For the most part, the subcultures, the countercultures, the dissenters, and the opponents continue mindlessly — or perhaps just helplessly — to follow the pattern of the dominant society in its extravagance, its wastefulness, its dependences, and its addictions.

The old problem remains: How do you get intelligence *out* of an institution or an organization?

My small community in Kentucky has lived and dwindled for a century at least under the influence of four kinds of organization: governments, corporations, schools, and churches — all of which are distant (either actually or in interest), centralized, and consequently abstract in their concerns.

Governments and corporations (except for employees) have no presence in our community at all, which is perhaps fortunate for us, but we nevertheless feel the indifference or the contempt of governments and corporations for such communities as ours.

We have had no school of our own for nearly thirty years. The school system takes our young people, prepares them for "the world of tomorrow," which it does not

NASA

Nobody can do anything to heal a planet. The suggestion that anybody could do so is preposterous.

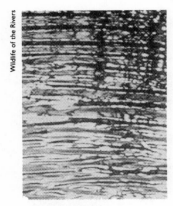

Wildlife of the Rivers

$O_{ur}$
understandable
wish to
preserve
the planet
must
somehow be
reduced to
the scale
of our
competence.

expect to take place in any rural area, and gives back expert (that is, extremely generalized) ideas.

The church is present in the town. We have two churches. But both have been used by their denominations, for half a century at least, to provide training and income for student ministers, who do not stay long enough even to become disillusioned.

For a long time, then, the minds that have most influenced our town have not been *of* the town, and so have not tried even to perceive, much less to honor, the good possibilities that are there. They have not wondered on what terms a good and conserving life might be lived there.

In this, my community is not unique, but is like almost every other neighborhood in our country and in the ''developed'' world.

The question that *must* be addressed, therefore, is not how to care for the planet, but how to care for each of the planet's millions of human and natural neighborhoods, each of its millions of small pieces and parcels of land, each one of which is in some precious and exciting way different from all the others.

Our understandable wish to preserve the planet must somehow be reduced to the scale of our competence — that is, to the wish to preserve all of its humble households and neighborhoods.

What can accomplish this reduction?

I will say again, without overweening hope, but with certainty nonetheless, that only love can do it. Only love can bring intelligence out of the institutions and organizations, where it aggrandizes itself, into the presence of the work that must be done.

Love is never abstract. It does not adhere to the universe or the planet or the nation or the institution or the profession, but to the singular sparrows of the street, the lilies of the field, ''the least of these my brethren.''

Love is not, by its own desire, heroic. It is heroic only when compelled to be. It exists by its willingness to be anonymous, humble, and unrewarded.

The older love becomes, the more clearly it understands its involvement in partiality, imperfection, suffering, and mortality. Even so, it longs for incarnation. It can live no longer by thinking.

And yet, to put on flesh and do the flesh's work, it must think.

In his essay on Kipling, George Orwell wrote: ''All left-wing parties in the highly industrialized countries are at bottom a sham, because they make it their business to fight against something which they do not really wish to destroy. They have internationalist aims, and at the same time they struggle to keep up a standard of life with which those aims are incompatible. We all live by robbing Asiatic coolies, and those of us who are 'enlightened' all maintain that those coolies ought to be set free; but our standard of living, and hence our 'enlightenment', demands that the robbery shall continue.'' *

This statement of Orwell's is clearly applicable to our situation now; all we need to do is change a few nouns: The religion and the environmentalism of the highly industrialized countries are at bottom a sham, because they make it their business to fight against something which they do not really wish to destroy. . . . We all live by robbing nature . . . but our standard of living . . . demands that the robbery shall continue.

We must achieve the character and acquire the skills to live much poorer than we do. It is either that or continue merely to think and talk about changes that we are inviting catastrophe to make.

---

* ''Rudyard Kipling,'' *A Collection of Essays by George Orwell*, Doubleday Anchor Books, 1954, pp. 126-127.

The great obstacle is simply this: the conviction that we cannot change because we are dependent upon what is wrong.

But that is the addict's excuse, and we know that it will not do.

How dependent, in fact, are we? How dependent are our neighborhoods and communities? How may our dependencies be reduced? To answer these questions will require better thoughts and better deeds too than we have been capable of so far.

I am not trying to mislead you, or myself, about the gravity of our situation. I think that we have hardly begun to grasp the seriousness of the mess we are in.

Our most serious problem, perhaps, is that we have become a nation of fantasists. We believe, apparently, in the infinite availability of finite resources. We persist in land use methods that reduce the potentially infinite power of soil fertility to a finite quantity — which we then proceed to waste as if it were an infinite quantity. We have an economy that depends, not upon the quality and quantity of necessary goods and services, but on the moods of a few stockbrokers. We believe that democratic freedom can be preserved by people ignorant of the history of democracy, and indifferent to the responsibilities of freedom.

Our leaders have been for many years as oblivious of the realities and dangers of their time as were George III and Lord North. They believe that the difference between war and peace is still the overriding political difference — when, in fact, the difference is diminished to the point of insignificance. How would you describe the difference between modern war and modern industry — between, say, strip mining and bombing, or between chemical warfare and chemical manufacturing? The difference seems to be only that in war the victimization of humans is directly intentional and in industry it is "accepted" as a "trade-off."

Were the catastrophes of Love Canal, Bhopal, Chernobyl, and the Exxon Valdez episodes of war or of peace? They were, in fact, peacetime acts of aggression, intentional to the extent that the risks were known and ignored.

We are involved everywhere in a war against the world, against our freedom, and indeed against our existence.

Our industrial accidents, so-called, should be looked upon as revenges of Nature. We forget that Nature is necessarily party to all our enterprises, and that she imposes conditions of her own.

Now she is plainly saying to us: "If you put the fates of whole communities or cities or regions or ecosystems at risk in single ships or factories or power plants, then I will furnish the drunk or the fool or the imbecile who will make the necessary small mistake."

And so, graduates, my advice to you is simply my hope for us all:

Beware the justice of Nature.

Understand that there can be no successful human economy apart from Nature, or in defiance of Nature.

Understand that no amount of education can overcome the innate limits of human intelligence and responsibility. We are not smart enough or conscious enough or alert enough to work responsibly on a gigantic scale.

Make a home. Help to make a community. Be loyal to what you have made.

Put the interest of the community first.

Love your neighbors — not the neighbors you pick out, but the ones you have.

Love this miraculous world that we did not make, that is a gift to us.

So far as you are able, make your lives independent of the industrial economy, which thrives by damage.

Find work, if you can, that does no damage. Enjoy your work. Work well. ∎

Secrets of the Old Growth Forest

**S**o far as you are able, make your lives independent of the industrial economy which thrives by damage.

# Healing The Dream of Apocalypse: A Ritual

*BY MICHAEL ORTIZ HILL*

*Illustrated by*
*PAUL MIROCHA*

---

*Restoration can involve a large public project or a small private act. Repairing the world out there provides the greatest value for us humans when it emanates from a healing of the heart within.*

*Michael Ortiz Hill lives in Santa Cruz, California, where he is at work on a book about dreams related to nuclear war called* The Dream at the End of the World. *—Richard Nilsen*

THE **B**OMB has been a member of my family — and also a figure in my dreams — since before I was born. My father grew up in Alamogordo, New Mexico, just south of where the first atomic weapon was tested at the Trinity site. His mother, a crusty Southern Baptist, was operating the switchboards for Ma Bell when the electromagnetic pulse snuffed out the lines of communication in the early morning of July 15, 1945.

My mother's family migrated to the northern outposts of New Spain up the Chihuahua Trail in the early 1700s — passing merely five miles from Trinity. In 1943, when Los Alamos, New Mexico, was only a sprinkling of lights after sunset on the nearby mountains, my mother would point them out to her little cousin from a ridge at my grandfather's ranch. ''That is where Santa Claus and his elves are preparing Christmas for us,'' she would tell him. After the war my parents lived in Los Alamos, where my older sister was born.

A couple of years ago I began writing a book on dreams related to nuclear war — both people's personal dreams that I am continuing to gather for research, and the collective or cultural ''dream'' that led to Trinity. What is this dream of destruction we seem unconsciously compelled to live out? What are the questions, the yearnings, that the Bomb attempts to answer for us? I want to know how we ''hold'' the Bomb in our deepest psyche. ''Who'' in us would destroy the world? ''Who'' would preserve and cherish it? What are these dreams asking of us?

June before last I made a pilgrimage to the Trinity site to perform a small ritual of ''healing the dream.'' I planted Kwan Yin, the Buddhist goddess of limitless compassion, in the earth at Ground Zero.

My mother, ten-year-old daughter Nicole, and her friend Lily dropped me off at the border of the White Sands Missile Range in the late afternoon. As I unloaded my backpack, Nicole sang to me, ''Shalom, Haverot, Shalom,'' — a Hebrew song for the one who departs on a journey or goes off to war.

I meditated in the tall grass with the little black ants and the grasshoppers until the sun set and the moon rose. I wanted to trespass under the cover of night. Then for six hours I walked alongside the north-south spine of the Sierra Oscura mountains.

The desert under the full moon was radiantly beautiful. To see an antelope walking slowly in the blue shadows or to be startled by the unearthly sounds of stray cattle, or even to come by something as mundane as a power line buzzing in the middle of nowhere — these things had terrifying presence.

My geological survey maps were all but useless, but it took little imagination to "see" the mushroom cloud burning in the night to the southwest of me. I found myself relying on this apparition and my compass to negotiate the twists and turns in the dusty roads.

Often I felt lost — in all possible senses of the word. In fact this "lostness" turned out to be an excruciating but essential part of my preparation for the ritual. By the time I got to my destination I had been utterly reduced to what was most elemental — often what was most fearful and confused in myself.

I sat there for two hours in a swirl of unbearable vulnerability. Paranoia of the military police; a five-year-old's sense of having trespassed — and both a fear of being caught by the "adults" and a hunger to be punished; flashes of dying a rather stupid and unnecessary death. I was aware of my complete incapacity to pull myself together or to surrender. I very seriously considered not doing what I had spent months preparing for — knowing well that to act falsely would be poison. Finally out of this incapacity and brokenheartedness I began with the prayer: "Make use of me. I am a little

child and will die a little child. I can't pretend otherwise. I do not know how to proceed. Lead me."

I stood up and paced out the four directions after the manner of the Mescalero Apache who used to come out to this land for vision quests before ranchers appropriated it in the late 1800s. Coyotes began howling shortly before dawn — first to the east of me and then to the west. I performed the ritual in a whispered voice. It was unexpectedly and exquisitely intimate in a way that I could never have imagined.

J. Robert Oppenheimer, who was a Sanskrit scholar as well as the "Father of the Atomic Bomb," wrote that when he first faced the mushroom cloud it was Vishnu's words to Arjuna that came to his mind: "Behold I am become Death, shatterer of worlds." In my original draft of the ritual I succumbed to the temptation of answering the pathos of apocalypse with an apocalyptic gesture — I planned a ritual act of burning the scriptures of apocalypse (the tenth chapter of the Bhagavad Gita and the end of the Book of Revelations) at Ground Zero. My wife, Deena Metzger, rather sensibly suggested that, instead of trying to be on equal terms with the Manhattan Project by replicating its grandiosity, perhaps tenderness and humility might bear the proper attitude of healing in such a dark period of history. So at the core of my rewritten ritual I called the "enemy" to the warmth of the heart. I made a small fire of twigs in the hole in which I would bury Kwan Yin, and prayed:

"Here where the nuclear fire first burned I make a hearth of a handful of twigs. To this fire I call all who I have feared and despised and felt superior towards; those people of my own dark dreams and those of my waking life who I have felt wounded or betrayed by: I invite you here without

## NEAR TRINITY SITE, THE DAY AFTER THE JUNE MOON WENT FULL

*When a swallow flies*
*toward the face of a cliff*
*its wings cut the air*
*with an effortless violence*

*And so it was*
*when the jets flew over,*
*the stunning grace as they curved*
*against the embankment*
*of the Sierra Oscura,*
*shuddering*
*along the spine of the yucca.*
*And beneath the roar*
*I also shuddered*
*with the dull grey beetles*
*that cluster*
*on the scat of coyotes*

*When the bombs began dropping*
*I thought, this cannot be*
*I thought*
        *El Salvador*
                *Afghanistan*
*and not long ago*
*a small hamlet in Vietnam*
*the fire, the wailing of mothers*
*over dead children.*

*There was no place to hide.*
*I became a dusty fetus*
*curled up amongst cactus*
*with only a small prayer*
*in a small voice:*

*"please if I die now*
*regard the life of my daughter with*
        *kindness*
*if she is to be fatherless*
*tend to her heart"*

*When the bombing paused*
*I stood up and walked hurriedly north*
*My back to where the mushroom cloud*
*first lifted poison to the sky.*
*The largest tiger swallowtail*
*I had ever seen*
*alighted on the ragged*
*blue flower of a thistle.*
*My God, this life*

*And then the bombs*
*began dropping again*

*After the bombs stopped the second time*
*an antelope looked up from its grazing,*
*held my gaze for a long moment*
*and ran off to where the earth still smoked*
*What must it think?*

*In 1945, two herd of antelope*
*scattered to these mountains*
*when the first nuclear bomb was*
        *tested here.*

*Later that day, J. Robert Oppenheimer,*
*a man not unfamiliar with tenderness,*
        *found*
*a turtle turned on its back near*
        *Ground Zero.*
*He set the turtle back on its feet.*

*Three weeks later, the Bomb*
*the Japanese would call the "Original*
        *Child"*
*leveled Hiroshima,*
*and then, Nagasaki*

*In this world*
*to frighten a butterfly*
*will never mean very much*
*To bake the underside of a slow reptile*
*or to shatter the minds of a herd of beasts*
*To burn to the ground*
*a whole city of children*
*has become the ordinary labor of*
        *ordinary men*

*Have mercy on us.*

                                        —MOH

demand but with an open heart so that I might look upon your face and speak with you. I may not be able to forgive and let go. I know that such things cannot be forced — but I desire to do so and will pray always for the courage to do so. When I am lost and my heart is numb I will try to remember this prayer — because it has always been true that each of you in your own way have been my most insistent and generous and difficult teachers. I hope that between now and the unknown time of my death I learn to cherish and carry to fruition at least a fragment of what you have shown me. I know that this is the way of lovingkindness which I am so naive in. The confusions of this world that would rely upon the Bomb and risk the life of everything that lives to be "protected" from the "enemy" — these confusions are very intimately my own as well. If making these vows at this place where a vast nightmare was born can be at all helpful for sentient beings I pray that it be so. If not — let it at least be helpful for my own suffering and towards those whom my life touches."

As the twigs burned to cinders I addressed my enemies one by one, informally. I recounted memories and tried to find the thread of empathy towards the personal dilemmas and pain that led them to be harsh or unconscious with me. I searched my own heart for the ways I had contributed to the circumstances in which I came to feel betrayed — and I requested the enemies' forgiveness. I reflected upon concrete gestures of reconciliation that might be appropriate — a phone call or letter, a meeting, a confession, an anonymous gift — or merely composting what I didn't know how to heal back into my dharma practice.

When I finished the sky was reddening with dawn. I buried Kwan Yin and kissed the earth. I knew I had to hurry north because my water supply was low and it promised to be a hot day. I also knew that the sun meant I was suddenly visible and ran a greater risk of being arrested.

Around 9:00 A.M. a bevy of jets flew over and began dropping bombs. This was the unexpected completion of my ritual. Whatever had eluded me at Trinity about the fragility and preciousness of life in this insane century became fully clear then.

Sometimes one is momentarily blessed with the ability to truly listen to what this planet, in her great distress, requests of us to offer towards the healing of this madness. It has been a year and a half since I did the ritual at Trinity. Since then I have come to feel that perhaps it is this ongoing labor of love, these efforts at reconciliation that give me the right to call myself a true citizen of the twentieth century.

The ritual at Trinity stands as a turning point in my life — the beginning of a slow and painstaking purification in the realm of my dreams and my everyday life that partakes of the most difficult vulnerabilities of my heart. To recognize that the enemy has a face no less human than my own — this has been the koan I have returned to again and again in order to realize self-acceptance and kindness toward others.

I have found that the enemy forces me to look at my own shadow — which is to say, forces me to recognize that I am not who I think I am. Much of the rage and fear that come up is related to being set adrift in unknown areas of my psyche — areas I have adamantly or complacently made a point of avoiding.

This was what happened when I got lost at the White Sands Missile Range on my way to Ground Zero. Staggering through the night in the field of the enemy it was his presence "everywhere" that set me adrift in the fears and insecurities I've carried with me since I was a child. As it was, it was these fears that devoured me, not the enemy himself — whom I never met. In other words the enemy I most feared, first "inner" but then, inevitably, "outer," bore the face of my own dark twin. Reconciliation, when it happens, has been the unexpected laughter that comes when I realize I was never other than a sibling to the ones I despised. ∎

## Bravo 20: The Bombing of the American West

*Photographer Richard Misrach has helped a generation to see American deserts with new eyes. Here he focuses on the pockmarked moonscape of a bombing range, located on public land in the high Nevada desert, and used illegally by Navy jets for over thirty years. The photos are mute testament to the awesome destructive power of modern armament. The text chronicles the battle by a handful of local citizens against the Navy, a small part of the struggle by rural Westerners against the enormous ongoing expansion plans of the U.S. military. Included is a proposal to reclaim the 64-square-mile bombing range into America's first environmental memorial — Bravo 20 National Park — complete with a visitor's center shaped like an ammo bunker and a "Boardwalk of the Bombs."*
—Richard Nilsen

**Bravo 20**
Richard Misrach and
Myriam Weisang Misrach
1990; 133 pp.

**$25.95** ($28.45 postpaid) from Johns Hopkins University Press, 701 W. 40th Street, Suite 275, Baltimore, MD 21211 (or Whole Earth Access)

●

Today's military is acting as though training done in the name of national security overrides both human rights and the laws of the United States. That has to stop. National security does not entail buzzing schoolbuses filled with children, scaring ranchers half to death, dumping fuel in wildlife refuges, or bombing historic vestiges. It does not place the military above the nation's environmental legislation, nor does it excuse toxic pollution and radiation contamination.

**Active eagle's nest.**

B Y 1953, when South Korea and North Korea signed the Armistice Agreement that ended the Korean War, the 151-mile-strip designated as the demilitarized zone had been devastated. Once littered with terraced rice fields, small crop plots and villages, the DMZ was bare of vegetative cover, pockmarked by bomb craters, and crisscrossed by hundreds of artillery roads. Towns and farms had been forsaken. Most of the region's forests had been razed to deny enemy North Koreans cover. Four million lives had been lost. So too had hundreds of acres of farmlands and wildlife habitat.

Today, the scarred slopes have been invaded by mixed hardwood forest. Terraced rice paddies have converted to marshland. Grass and shrubs have conquered abandoned farms. The trumpeting cries of the endangered Manchurian cranes replace the sound of gunfire; pheasants, deer, lynx, and occasional tigers roam the still heavily land-mined area. What is a military no-man's land — site of the longest cease-fire in history,

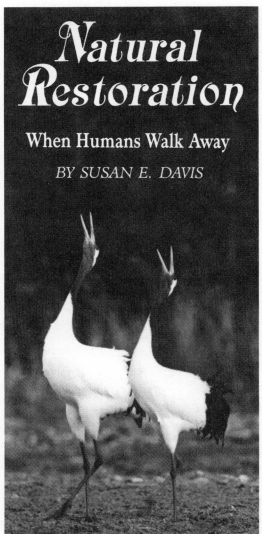

# Natural Restoration
## When Humans Walk Away
### BY SUSAN E. DAVIS

*It is damn near impossible for us humans to step out of the picture, and harder still to realize when we do that Nature doesn't miss us.*

*Susan Davis is a free-lance writer who lives in Berkeley, California.*
*—Richard Nilsen*

**Endangered Manchurian cranes find sanctuary in the abandoned and still heavily land-mined Korean DMZ.**

International Crane Foundation

only one gunshot from explosive conflict — has become a wildlife sanctuary. And it has become that with no aid from humans — other than removing our presence.

The Korean DMZ illustrates an idea many of us forget: Nature heals itself. In the midst of warnings of humans' irrevocable impact, news of an irrepressible nature sounds peculiar. But the DMZ story is not an isolated example. Plants and animals are very stubborn. Given a disturbance, an ecosystem regenerates, even without our helping hands. Put another way, ecosystems regenerate if we leave them alone.

What's curious is that we've forgotten that.

If you ask most people to think of places in the world that "came back" after devastating human impact, you'll stump them. Most pause and say, "What an interesting question." Many look nervous. The idea does not fit environmentalist, or anti-corporate, thinking. It's easier to think of areas that humans ruined than it is to think of areas that nature healed.

If you ask a professional ecologist if nature regenerates, you'll get a different answer. Ecologists are different from environmentalists. "Ecology" connotes *study* — generally of ecosystems. "Environmentalism" connotes *advocacy* — also of ecosystems. But an ecologist won't grimace at

the idea that nature comes back. An ecologist will ask "What do you mean by 'come back'?"

Good point. Recovery can refer to vegetative cover which differs from pre-disturbance cover. Swordgrass and shrubs, for example, cover much of the 9,000 acres the U.S. bulldozed in Vietnam during the war. But that area used to be hardwood forest. Trees grow on Enewetok, Pacific test site of nearly 50 U.S. nuclear bombs in the 1950s and 1960s. But the few inhabitants living there cannot eat the fruit because it is so highly contaminated with radioactivity. "Nature will always adapt," says Neo Martinez, an ecologist with UC Berkeley's Energy and Resources Department. "The question is with what."

In more specific terms, however, recovery refers to an ecosystem returning to its pre-disturbance condition. Korea is a good example of this kind of recovery. The DMZ returned to its condition before the impact of 20th-century human war and before the impact of human agriculture — which began several thousand years ago.

Whether or not a devastated area recovers depends on a number of conditions. Most fundamentally, the site needs its original topsoil. That top layer, also called the top, or A, horizon, contains the nutrients and microfauna necessary for plants to grow. Its composition may result from thousands of years of geological processes. "When you work in ecology long enough, you become obsessed with soils," says Peter Warshall, an international biological consultant. "I almost think about soils more often than I think about plants."

In temperate zones, topsoil often contains seeds, and stem and root fragments, from which new plant life emerges. Some seeds can lie dormant in the topsoil for hundreds of years. But if that A horizon is removed — in the course of strip-mining, or erosion from clear-cutting — there is, literally, no groundwork from which an ecosystem can develop.

Similarly, if the top horizon is altered — by adding chemicals, ploughing or planting crops — a different kind of vegetation emerges when the site is abandoned. That's what happened on many now-abandoned farm sites in New England and the Midwest. But if neither fertilizers nor crops are introduced, even heavily used areas can return to the condition of nearby undisturbed areas. Certain grazing lands in Tanzania's Serengeti National Park, for instance, were cordoned off from domesticated animals, and recovered within two or three years.

Even when the topsoil of an area is not removed, full recovery will not occur without seeds. Some lie dormant in the soil. Others drift in from nearby vegetated areas. Birds and animals also carry seeds in their feathers, fur or digestive systems. Damaged ecosystems surrounded by undamaged areas have an easier time recovering than those isolated by geographical barriers, such as oceans, mountains, deserts and cities. And once deposited, the seed has to fall into a niche that facilitates germination. Sunlight, moisture, temperature and nutrients all have to be in the right proportions.

The process of the subsequent vegetative recovery is called "ecological succession." Its first proponent was naturalist Frederic C. Clements, who claimed that all ecological formations follow an orderly progression from inception to perpetually stable "climax communities," or self-replicating assemblages of plants. "As an organism," he wrote in 1916, "the formation arises, grows, matures, and dies. . . . The climax formation is the adult organism, of which all initial and medial stages are but stages of development."

Clements's theory was strongly rooted in Darwinian ideas of biological evolution, and spin-off ideals of social "progress." Today, most ecologists don't believe in "climax communities." Ecosystems change continually, they say. Targeting an ideal, or even stable, formation is next to impossible. The system may pass through a "mature" or "climax" state, but it may be on the way to something else, often to something simpler.

Still, most ecologists recognize that succession occurs. Hardy "pioneer" species invade a disturbed site, and are then replaced by a series of other plant, insect, and animal associations. The progression, for many modern-day ecologists, is not towards stability, but within constantly changing factors.

Recovery also depends on the stability of an ecosystem. Stability, according to ecologist C.S. Holling, is a system's ability to absorb change and recover rapidly. (Another ecological term, "resilience," refers to a system's ability to withstand disturbance in the first place). Arid regions are generally less stable than moist ones. You can still see tank tracks, for instance, in the Southern California desert, where General Patton led his Desert Training Corps — in WWII. Such marks would disappear in the rain, litter and vegetation of a northern forest.

No matter what the stability of an ecosystem, however, recovery really depends on time and space. That is, if the time-scale is long enough, and the space-scale is small enough, most areas will revegetate, given the appropriate seed and soil conditions, and an appropriate "hands-off" attitude by humans.

## RAINFORESTS

For centuries, humans in rainforests in Central and South America, Asia, and Africa have relied on natural restoration in plant cultivation. In slash-and-burn agriculture, forest dwellers chop 1- to 3-acre plots, and then burn the "slash." Unlike temperate forests, 70 percent of a rainforest's calcium, magnesium and potassium are stored in the biomass above ground. Burning releases nutrients into the soil. Farmers cultivate plants for two or three years, until the soil wears out. Then the farmers move to other plots, allowing the original site to grow back.

Christopher Uhl, an ecologist at Penn State, studies whether or not rainforests can come back after disturbance. Uhl has found that despite the trauma of slashing and burning forest areas, cultivated plots can return to rainforest within 100 years — if they're left alone. In the first year, seeds for woody pioneer species blow in from plants surrounding the plot. Often they germinate in the safe shade of fallen logs or random pineapple trees, where dead leaves and nutrients nurture the seedling. These pioneers then create the shady micro-climates that then enable slower growing, primary forest species to germinate.

Small plots and long fallow periods allow tropical forests to regenerate. But when a plot is too large, it cannot regenerate as easily. Uhl studied a 25-acre site near San Carlos, Panama, which had been cut and bulldozed in 1971 to build a military post. Two years later, the post was abandoned, leaving all 25 acres deprived of regenerating mechanisms. The bulldozing scraped off the topsoil, with its precious seeds, stems and roots. It also removed the felled biomass, thereby stripping the site of nutrients and safe germination sites for seeds from nearby forest.

Uhl predicts that the 15-acre central plot will remain totally bare for a millennium, because it is too far from the nearest seed trees. Perimeter acres show some signs of recovery, in the form of grasses and shrubs. But none are taller than 10 feet. This area too will take centuries to revegetate. The danger here is that certain species can be permanently lost if their delicate habitat does not regenerate. ►

That's only 25 acres. One thing that Uhl's studies — and those of others, including Robert Bushchbaker of the Conservation Fund — highlight is that devastated areas need proximity to non-disturbed areas in order to regenerate. Currently 150 acres of rainforest fall per minute in the Amazon. 3.6 million acres are cleared each year. Fifty-five percent of the original tree cover is already lost. Miles-wide swatches have been stripped of vegetation, burned repeatedly and packed down by cattle hooves. Those areas won't regenerate — and their hot, open spaces will damage remaining stands of forest.

## TEMPERATE FORESTS

''Trees,'' wrote Steward H. Holbrook, in the 19th-century *Yankee Loggers*, ''merely hide in the soil until the man with the axe or hoe has turned his back for a moment, instantly to send up their first shoots to advance in astonishing numbers and size until all signs of man's efforts have been obliterated.''

Holbrook may have been exaggerating, but northern temperate forests do heal more easily than rainforests, even after massive clearcutting takes place. Warm, wet conditions facilitate germination and rapid growth in rainforests. But northern forests have less than one-tenth the number of species per acre than do rainforests. And that — in combination with a deep bed of vegetative litter, and highly resilient seeds — means that northern forests have an easier time returning to their condition pre-disturbance.

The Shenandoah National Park is ''probably the largest-scale recovery from human abuse we know,'' says Duncan Morrow, Recreational Resource Assistant of the National Park Service. The Shenandoah had been farmed, grazed, mined, burned and logged for over 200 years before being made a national park in 1935. At that time, one-third of its 195,000 acres was open grazing land. All but 5 percent had been cut over at one time or another.

Fifty years later, hardwood deciduous forest covers most of the area. Containing over 108 tree species, the forest is one of the most diverse in the world. Oak and hickory predominate, with sprinklings of American beech, sycamores, and tall tulip poplar. The undergrowth is heavy with mountain laurels, ferns, bushes and shrubs. Deer, bear, fox, bobcat and cougar roam throughout. Roads are crumbling, and abandoned houses are crawling with vegetation. Besides those remnants of human presence, ''it's hard to distinguish between these areas and deciduous areas that have not have been clearcut,'' says Chuck Anibal, chief interpreter for the park. ''There is no obvious aesthetic difference to the casual observer.''

Where there is a difference is between the secondary oak forest and the virgin pockets of densely packed 200-year-old eastern hemlocks. There, thick canopy shades dark, open forest floor. But the qualitative difference, Anibal insists, is due to species composition, not time.

In 1976, Congress added 125 square miles to the park, under the Wilderness Preservation System. When the Wilderness Act was passed in 1964 it made the U.S. the first country to legislate a recognition of the value of ''wilderness.'' The act defines ''wilderness'' as ''an area where the earth and its communities of life are untrammeled by man.''

The crucial elements in the Shenandoah recovery were that it was left alone, and that sufficient seeds sources existed for regeneration. As in the rainforests, when human impact ceased, nature had a chance to heal. But unlike rainforests, the Shenandoah didn't lose its seed sources. Seeds still existed in the soil, and the forest was never completely clearcut. Uncut, diverse segments still existed.

The same is true with the demilitarized zone in Korea. The armistice signing ended the shooting phase of what is now called the Forgotten War. It also ended any civilian presence in the area. Entrance into the DMZ is strictly controlled now. Only that has allowed the 2.4 mile-wide zone to regenerate — with some of the tallest patches of wild forest in Korea — and to create feeding grounds for the endangered white-naped and Manchurian cranes.

Had the DMZ been ''reforested,'' the tree cover would be a stiff-rowed monocultural forest. Areas in Korea that were reforested after the war still look ''scalped,'' says George Archibald, a biologist with the International Crane Foundation, who has done significant work in creating Korean preserves for endangered cranes. Instead, the DMZ hosts a mixed arboreal regrowth.

''It took my breath away,'' says Bernard Trainor, military correspondent for *The New York Times*, who fought in the war and visited the DMZ last year. ''You'd never know it had been turned to sand during the war. I was able to identify all of the terrain features, but it looked com-

Human habitats and structures are allowed to slowly disintegrate in the Shenandoah National Park, as part of a hands-off policy.

National Park Service

Now a lush wetlands, this naturally restored Minnesota quarry provides a habitat for nearly twice as many bird species as the same area did before mining.

Dan Svedarsky

pletely different. Now it is overgrown with foliage. It was a very emotional experience.''

Nature often does a better job at restoration than humans do. Take strip mining, for instance. Appalachian sites mined thirty years ago, before reclamation regulations, have begun to regenerate on their own. ''Nature has done a remarkable job so far,'' says Skip Deegan, coordinator of the Sierra Club's Strip Mining Project. ''Now we primarily have nitrogen-fixing trees, like locust, and autumn olive. But as you go further down the mountain, you see little hardwood trees beginning to spring up. Within 20 to 25 years, hardwoods will return.''

Mining companies today have to reclaim strip sites to their original condition, or ''better.'' Restored areas often contain only ''drought-resistant grasses that pop up within a day or two,'' Deegan says. These are supposed to increase the value of the land. ''Basically it's stabilized,'' Deegan says. ''There's some erosion, but nothing too outrageous. But this is a long shot from a good solid growth of trees.''

Similarly, abandoned, mined-out gravel pits can sometimes regenerate more successfully than those that are ''restored.'' Before mining began, at what is now called the Red River Natural History Area, near Crookston, Minnesota, the area was mostly prairie lands. When the 85-acre site was in operation, it was dusty, barren, and open. When mining ceased in 1960, the pits had filled with groundwater, which eventually created wetlands. Willow and cottonwood seeds floated in on the wind and germinated in the damp soil. Seeds of flower-bearing shrubs like snowberry and raspberry came next. Today, the area is lush with clusters of cottonwood, dogwood, and cattails, honeysuckle, chokecherry, and nannyberry. Red-winged blackbirds, warbling vireos, skunks, foxes, and even moose, use this as habitat.

''Many of the areas mined and then left behind could not have been wildlife habitat if they had been reclaimed according to regulations,'' says Dan Svedarsky, head researcher at the site, which is run through the Northwest Agricultural Experiment Station, at the University of Minnesota. ''They would have been used for agriculture or development.''

For some wildlife, Red River regenerated to a condition better than before mining. Today, nearly twice as many species of birds use the site as habitat than were recorded when mining began in 1900. Thirty-eight of those species are rare or endangered.

Both the DMZ and the gravel pit instances raise a ''what next?'' question. Regenerated areas tend to attract humans. Scientists in Great Britain have understood the habitat value of restored gravel pits since the 1930s and 1940s, when they found that the wetlands attracted rare birds, including the great crested grebe and the little ringed plover. Gravel companies quickly converted habitat value to recreational value. Many sites now offer fishing, swimming and power-boating. One site, operated by Ready Mix Concrete Ltd. even offers a Disney-like theme park. That may not attract wildlife.

Unless naturally restored areas are protected, humans will return. Red River and the Shenandoah National Park are both protected. But if North and South Korea ever reunite, human politics could wreak havoc on the present environment of the DMZ, just as it inadvertently allowed a wildlife sanctuary to emerge there 36 years ago.

One interesting thing about ecology is that as you learn, you see. When I drive down city streets now, I marvel less at human omnipresence — walls of brick and carpets of concrete — than I do at the sheer persistence of non-human life. Weeds struggle through pavement, saplings sway in sidewalk niches, dandelion puffs sail past plate glass windows, seeking soil in which to lodge. Some of these travelers look tenacious, healthy, resilient. Others look neurasthenetic. But they're there.

Some human impacts are undeniably irrevocable, especially those that persist in time and space long after the actual activity ceases. DDT disperses through the food web, thereby affecting plants and animals thousands of miles and decades from the original spraying. Plutonium has a radioactive half-life of 24,000 years. Chlorofluorocarbons destroy ozone molecules 150 years after their release in the atmosphere. And every extinction is forever. If anything, understanding nature's ability to heal should drive us to more adamantly regulate persistent activities and substances, to allow damaged areas to recover, and to prevent further losses. ∎

## The Control of Nature

*Although John McPhee describes this book as "a planned vacation from projects in geology," his fascination with the forms earth takes, especially plate tectonics, continues unabated. This is fortunate for the reader, because he is one of the best explainers writing in English. This time he presents Geology With a Human Face. Here are people versus nature in extremis — keeping the Mississippi River from jumping its banks and bypassing New Orleans, fighting a lava flow threatening a town in Iceland, and playing catch with the flood debris that washes out of the San Gabriel Mountains and into the streets of Los Angeles. His leisurely assemblage of fact and anecdote examines the lives of people caught in desperate environmental situations, and demonstrates that the difference between "rational" and "rationalize" often depends on where you live.*

—Richard Nilsen

### The Control of Nature

John McPhee, 1989; 272 pp.

**$8.95** ($11.70 postpaid) from Putnam Publishing Group/Order Dept., P. O. Box 506, East Rutherford, NJ 07073; 800/631-8571
(or Whole Earth Access)

•

"Those people are crazy." He means that Pasadena Glen has the compact dimensions of a bobsled run, and the disassembling mountains hover above it. He means that from time to time all hell will break forth from the mountains. . . . The glen is so narrow that its houses are perched between streambanks and canyon walls. The nearest debris basin is below them, and therefore not meant to help them. . . .

"People come in and live — as we do — where we really shouldn't live," Mel Horton remarked to me, tendering the explanation that periods between serious floods are often long.

"It's a fantastic place to be in a storm," his wife, Barbara, said. "You hear a sound like giant castanets — boulders clicking together. They're not pebbles. And there is a scent, which is absolutely heavenly, of the crushed chaparral plants. It's so fragrant and beautiful it's eerie to have it associated with something so terrifying. And, God knows, it is terrifying."

"So why do you live here?"

"Freedom," Mel said.

## To Govern Evolution
## • The End of Nature

*The subject matter of these two books will directly and indirectly dominate political dialogue far into the future. Both authors agree on at least one point: humanity has irrevocably crossed a line into a new era for the planet. The entry into this new era is unwittingly of our own doing and will have profound consequences for all life. From that point Walter Anderson and Bill McKibben diverge. Dr. Anderson, a political scientist, optimistically believes that with the right political control mankind can still make the most of "shooting the genetic rapids." McKibben, a nature writer, lays out a more detailed case for why, regardless of the decisions we make, we are all in for a dunking in those rapids.*

*McKibben splits our choice of paths to the future into the "defiant" and the "humble." Anderson clearly opts for the defiant, that is, to continue dominating all nature for our own ends, but to get better at it. The humble path, on the other hand, means recognizing that nature (or tattered remnants of it) has a right to exist for its own sake. Rather than using our technological prowess to accommodate an ever-growing human population, the humble path advocates controlling population, undoing past mistakes, and stabilizing global climate. McKibben's heart seems to be with the humble path, but his lifestyle, as he admits and Anderson would insist, is locked into the defiant path along with the rest of us.*

*Anderson explicitly advocates some form of world government as the way out of our problems, coupled with greater environmental education of the electorate. His optimism that the fossil-fuel-generated lightning bolt of human expansion can continue into the foreseeable future has an aspect of religious faith that technology and the wisdom to make the right choices will see us through. Like religion, the catechism of No Environmental Limits can be used to soothe our fears and justify our abuses. But even McKibben, with his pessimistic grasp of the dire straits we are in, understands that technology offers our only hope of softening the impact as we career into environmental limits. Both books are worth reading to keep up on the limits/no-limits debate.*   —Tom Ness

•

We have taken politics to consist of interactions among human beings. . . . Evolution, on the other hand, has to do with nature, with plants and animals and their environments, with a grand sweep of change that proceeds according to laws beyond human reach. That is how we think of the two — and that is the problem, because the two are now one. They have flowed together, and there is no making sense of one without reference to the other.

### To Govern Evolution

Walter Truett Anderson, 1987; 376 pp.

**$22.95** ($24.95 postpaid) from Harcourt Brace Jovanovitch, 465 S. Lincoln Drive, Troy, MO 63379; 800/543-1918
(or Whole Earth Access)

### The End of Nature

Bill McKibben, 1989; 226 pp.

**$9.95** ($12.45 postpaid) from Doubleday & Co./Cash Sales, P. O. Box 5071, Des Plaines, IL 60017-5071; 800/223-6834
(or Whole Earth Access)

The human species has, in a burst of creativity over the past forty years or so, transformed its institutions of governance and brought forth the first world order, more or less behind its own back.

This is a pretty rickety system, and any number of things may happen: It may collapse and take the biosphere with it. It may turn into a much more unified, even tyrannical, world government. It cannot, given the present inertial roll of globalizing forces, return to a system of autonomous nation-states. Neither can it, for the same set of reasons, sort itself out into a system of bioregional ecotopias. The latter is an attractive scenario, but not achievable this side of the population wave, since people have a tendency to move into your bioregion if they happen to be starving in their own.   —To Govern Evolution

•

The idea that the rest of creation might count for as much as we do is spectacularly foreign, even to most environmentalists. The ecological movement has always had its greatest success in convincing people that we are threatened by some looming problem — or, if we are not threatened directly, then some creature that we find appealing, such as the seal or the whale or the songbird. The tropical rain forests must be saved because they contain millions of species of plants that may have medical uses — that was the single most common argument against tropical deforestation until it was replaced by the greenhouse effect. Even the American wilderness movement, in some ways a radical crusade, has argued for wilderness largely as places for man — places big enough for backpackers to lose themselves in and for stressed city dwellers to find themselves.

But what if we began to believe in the rain forest *for its own sake?*

—The End of Nature

# 2.

# Environmental Restoration In The USA

ANY PLACE HUMANS have changed the natural world there is likely to be a need for environmental restoration. Repair work is going on today in an amazing variety of locations. Steve Packard begins this section with a look at the suburbs of Chicago, where volunteers are busy solving The Case of the Missing Ecosystem, in thickets that are within earshot of a freeway's dull roar. On one of the most remote sections of the California coast, Freeman House chronicles restoration that began after noticing that one key species — salmon — was disappearing. This work has grown into a community-wide effort encompassing an entire watershed. At Redwood National Park and near Everglades National Park, it is governments that are hefting restoration shovels to undo the mistakes of the past. Mike Helm and George Tukel apply the lessons of ecology to cities, and propose the restoration of urban public spaces. They argue that the same rules that allow diverse natural habitats to evolve can also be used to fashion urban areas that are safer and more interesting to live in. David Sucher reports on a program under way in Seattle that is encouraging good urban design by recognizing it with public awards. All of these efforts involve careful attention to detail — a prerequisite for the art of restoration in any environment. Among the smallest details are seeds, especially the ones that grow into mighty trees. Tree planting has become a popular solution to a Big Problem — global warming — a way to soak up some of the carbon our lifestyles continually spew into the atmosphere. Richard Sassaman describes the long tradition of tree planting in these United States, and Marylee Guinon points out that tree planting can — in its current manifestation as a politically correct green fad — sometimes be more of a problem than a solution. The art, and the forests of the future, are in the details.

Dixon Telegraph

# Just A Few Oddball Species:

## Restoration And The Rediscovery Of The Tallgrass Savanna

### BY STEVE PACKARD

I T WAS NOT OUR INTENTION to rediscover a lost eco-system. We were trying something else — to restore a tall-grass prairie landscape in an aggressive, non-compromising way. The North Branch Prairie Project now involves scientists from eight institutions and hundreds of volunteer workers. But when we began in 1977, we were just a few individuals with ideas and determination. Our initial goal had been to nurse back to health some small degraded prairies that survived in the Cook County forest preserves along the North Branch of the Chicago River, beginning within the city of Chicago and extending just beyond its borders in the northern suburbs. Most of our seven sites were at first little more than small openings in thickets of young brush. The grassy, flowery, open portions ranged in size from a few acres to a small part of an acre. Our idea was to enlarge these remnants by clearing the brush around them and planting prairie species in its place. We gathered seeds for many of the rarer things from the rapidly vanishing original prairie patches along railroads and other odd spots. The brush we cut consisted

*This tale of successful restoration combines enthusiasm, including a willingness to break the rules, with erudition; and has the elements of a good detective story. An earlier version appeared in* Restoration & Management Notes *(see page 76).*

*Steve Packard is Director of Science and Stewardship in Illinois for The Nature Conservancy, a mainstream preservation group that increasingly finds itself involved with restoration projects. For more information about the prairies and people in this article, contact TNC, 79 West Monroe, Chicago, IL 60603. —Richard Nilsen*

*VIRGINIA WILD RYE*

(Left) A volunteer sows seed to restore grasslands. The seed is a dry mesic prairie mix of about 50 different species, one of 12 basic mixes developed for different types of prairie and savanna. In all, more than 150 species are used.

•

mainly of non-native European buckthorn *(Rhamnus cathartica)* and large numbers of ash, elm, cottonwood, and other native trees. Visitors to these public forest preserves and our own Sierra Club volunteers sometimes questioned — and sometimes vigorously challenged — the cutting of all these trees.

Our objective was clear, however. It was to restore these tracts to their original, natural condition. The pre-settlement government survey of the area had recorded open prairie, with patches of oaks mostly along the river. We impressed on our critics that these prairie relics were extremely rare, while young brushland makes up a major part of the Forest Preserve District's 67,000 acres. Thorn thickets too dense to walk through in places represent not only a loss of habitat for the native animals and plants, but also a loss of useable recreation land for the public. These formerly open prairies are worth billions of dollars today, yet they exist as an impenetrable urban wilderness, increasingly at risk from development as they degrade.

How can we expect to maintain the political constituency to protect and expand these preserves if they degenerate into overgrown thickets?

Tallgrass savanna was deemed worthy of preservation earlier in this century for the same reason it makes a first-rate recreational landscape today — it was a natural, open ''parkland.'' This happy coincidence between what ecologists should want for biological conservation and what the general public would want for recreation and aesthetic purposes is an important factor in the ongoing restoration.

People responded quickly to the purity and grandeur of the vision. Right in the metropolis we would restore something of real cultural and ecological significance.

Though surrounded by a heavy growth of brush, some of the old oaks remained, and there was debate about what to do in the areas under those oaks. Did we envision forest? Or savanna? How close to the trees would we let our fires get when we burned the prairies? Particularly in areas rich in spring woodland flora, there was strong sentiment for protecting the ''forest.'' But another element of the vision compelled us to attempt to restore what

we had read of so often, something that no longer existed anywhere — the rich grassland running up to, and under, and through the oaks. The tallgrass savanna — a prairie with trees.

The question was, how do we bring the prairie together with the existing old oaks? Along the North Branch, the bank of thicket that separated our prairies from the ancient oaks ranged in width from 50 feet to a quarter mile or more. Where the oaks were close, we could see their heavy, twisted black limbs through the thin gray branches of ash and aspen. It was as if they were in prison or refugee camps, their lower limbs dead or dying in the deepening shade. ''Free the oaks!'' we sometimes joked. We felt the prairie and the natural woodland were lonely for each other, incomplete and unhealthy.

We wanted to use natural forces to bring them together, however. So we relied on the fires we were using to restore and maintain the adjacent, open prairies. We let the fires blast into the brush lines as far as they would go. ''Let the fire decide,'' became our motto. That had been the natural scheme; that's what we wanted. But most years, in most cases, the brush patches wouldn't burn. Ten-foot grass flames would sear the outer edges. But in the thicket where there was no grass, just green wood and matted leaves, the blaze quickly dwindled and flickered out. Most of the brush that did burn quickly grew back. We began to believe that the question was, did we have enough determination and patience to give natural processes two

Steve Packard

(Right) An ancient bur oak with lower limbs and understory grasses, flowers and animals shaded out, disappearing, or gone, as brush invades in the absence of fire. Buffaloes and Indians probably walked under this tree — bur oaks live up to 400 years.

or three hundred years to work themselves out? Or could we find something quicker?

## Interesting Failures

The obvious way to speed up the process of pushing the prairies back under the oaks was to continue clearing brush from the edge of the prairie right up to the trees. So, on April 5, 1980, six of us cut and dragged away the brush that separated the west edge of our Indigo Prairie from a patch of large bur and Hill's oaks. Onto the friable black earth under the oaks, where 5-inch DBH [diameter at breast height] cherry, hawthorn, and elm trees had stood, and out into the open, we broadcast about half a bushel of our choicest prairie seed mix. Then we raked it in for a comfortable start on its historic mission.

Two years later, when we evaluated the results, the prospects seemed hopeful. Thick stands of switch, Indian, and Canada wild rye grasses stood shoulder-to-shoulder with rattlesnake master and yellow coneflower; down on the ground the neatly sawed stumps of the once-invading trees were sprouting mushrooms. For some reason the grasses hadn't taken so well at the west edge, closest to the trees. But at least we had a natural ecotone, and nature could now take its course.

Buoyed by our apparent success, we cut the brush away from many oak edges and planted much more prairie seed. By the third year, however, it was clear something was wrong. Near the trees a few species, notably Kalm's brome *(Bromus kalmii)*, yellow coneflower *(Ratibida pinnata)*, and wild bergamot *(Monarda fistulosa)* did tolerably well at first, but the plantings remained thin, produced little fuel, and burned poorly. Gradually they filled in with brush. The leaves of the brush species quickly matted and rotted into the soil; the fires couldn't burn across bare dirt. And even under the oaks, where our fires burned well enough through the crisp oak leaves, none of our seed grew. What was wrong?

## Buckthorn Thicket

Heartsick at the depredations of lumbermen and sheep, John Muir once lamented that the ground layer of the Sierra forests, "once divinely beautiful, is desolate and repulsive, like a face ravaged by disease." The prairie groves of Illinois were once as grand and flowery as the natural "gardens" Muir championed in the western mountains. In Illinois, however, overgrazing is just half the problem. The other half is protection from fire. Without fire, the prairie groves sicken and deteriorate. The most obvious symptoms of this deterioration are infestations of European buckthorn, Tartarian honeysuckle *(Lonica tatarica)* and garlic mustard *(Alliaria officinalis)*. These aliens create thickets so dense, green up so early in spring, and hang on so late in fall, that they often drive out everything else. An especially sad (and common) landscape features forlorn, aristocratic old oaks in an unbroken sea of buckthorn — the understory kept so dark by the alien shrubbery that not one young oak, not one spring trillium, not one native grass can be found. Except for

PRAIRIE CONEFLOWER

Pasture and Range Plants

I N Illinois, overgrazing is just half of the problem. The other half is protection from fire.

the relic oaks, whose decades are numbered, the original community is dead.

Early publications of the Forest Preserve District, printed in the 1920s, show gracious open groves with tablecloths spread underneath for picnics. To traverse some of the same ground today would require an armored vehicle, or dynamite. Tackling it with merely a machete would be slow going. Don't wear your goose-down parka. It would soon be shredded and plucked. In some places you can explore the preserves only by crawling for long stretches on bare dirt under the dead, thorny lower branches of buckthorn.

The remarkably farsighted 1913 law authorizing the District calls for it "to *restore* (emphasis added), restock, protect and preserve the natural forests and said lands (i.e. lands connecting such forests), together with their flora and fauna, as nearly as may be, in their natural state and condition, for the purpose of the education, pleasure and recreation of the public." That noble statement — and the courage of District staff people in supporting our experimental work — emboldened us. Frustrated by the failure of our attempts to expand the prairie into the brushy areas near and beneath the oaks, we decided to leapfrog the persistent brushy border and to recut our fire lines to bring the fire in behind the brush, into the heart of the woodlands themselves. That would mean an unburnable strip of pure brush would still separate the grassy prairie from the oak-leafy grove. But at least the fire would be attacking the brush from both sides.

(Above) Americans have learned fire ecology from the lumber interests (Smokey the Bear) and Walt Disney (Bambi). The principal reason for the loss of prairie and savanna on ''protected'' conservation lands has been fire suppression. This sign, along a neatly mowed roadside, serves almost as a tombstone for the largely vanished midwestern prairie and savanna species that once comprised the 1,500-acre nature preserve behind it.

(Right) Volunteers burning prairie on the North Branch in Cook County. In Illinois, there are no large rural prairie preserves on good soil; they are *all* farmland. The only remaining specimens are in and around cities. Prairies must be burned to be sustainable, and educating nearby homeowners of this fact is a necessary step. Once neighbors understand the process and see the beauty of the prairies that result, they become supportive.

## First Burns

Our first burn deep in a bur oak grove was in the spring of 1984. It led to a comedy of soaring and dashed hopes. We were profoundly relieved when this ''forest fire'' went without mishap. We were depressed that the mostly 4- to 5-inch flames gave every appearance of being ineffectual. (Occasional patches of denser fuel flared up mightily.) But we soon found to our delight that even the little flames top-killed most of the buckthorn and Tartarian honeysuckle in the grove. Aside from small resprouts, the understory had been transformed from dense brush to entirely open; where visibility had extended just five or ten feet, you could now see for 50 to 100 yards.

We crawled on the bare dirt that summer, expecting prairie plants to appear, and looking, as always after burning a new area, for a few surprises. It was an exciting time. But the excitement gradually dimmed when nothing much appeared. Even after a second, much hotter fire and a second summer, most of the soil remained bare, waiting.

It's been 150 years since the natural savanna flourished in this grove, though; the seeds may not last that long.

There was reason to suspect that the problem was not a defunct seed bank of residual prairie species. In a few places we had actually sown prairie seed, choosing spots where we or the fires had cleared brush and left a hole in the canopy. Such planting had resulted in diverse, dense stands of prairie plants out in the open, but among the oaks we got at most a few wispy, flaccid shoots. In hospital nurseries doctors use the term ''failure to thrive'' to describe babies with a puzzling, sometimes fatal, wasting syndrome. Our seeded ''prairie understory'' was failing to thrive.

And a parallel ominous development gradually loomed up. During that second summer my dirty-kneed search for unfamiliar seedlings turned up increasing numbers of weeds. Canada and bull thistle, dandelion, briars *(Rubus sp)*, and burdock *(Arctium minus)* were increasing exponentially. Not forever was this bare ground going to

Pasture and Range Plants

*INDIAN GRASS*

# P
ART of our problem was that we were thinking too much about prairie and weren't picking up what this other community — the savanna — was trying to tell us.

wait for a rare community to find it. My judgment was that within two seasons the handsome galleries the fire had opened up under the oaks would become a dense tangle of mostly alien weeds. Conservative native species might conceivably out-compete them over the long run, if any emerged without the ''failure to thrive'' syndrome. But that process might take a very long time once a dense briar and thistle patch had become established. Would anyone go through the annual difficulties of getting this site burned decade after decade if it didn't seem as though we were on the right track? Would I? Unless something changed quickly, it looked as though we'd soon have an unsuccessful experiment on our hands.

Searching for an explanation, I visited the oak edge plantings at some of our other sites. At Indigo, the prairie grasses still stopped just short of the oaks. A dense stand of wild bergamot, yellow coneflower, and Canada wild rye *(Elymus canadensis)* ventured a little closer to the trees than the bluestem and Indian grass. But the prairie species just wouldn't move in under the oaks. In the shade behind the bergamot was a zone of alien thistles and bare dirt. I pulled the thistles, as I had a couple times a season for three years. But it was clear that they were gaining on me. And similar things were going on at a number of our other sites.

At Miami Woods, on the other hand, we initially thought the plantings were doing better. Plenty of grass appeared. But the grass, when it headed out, wasn't what we'd sowed at all. Instead it was three species unfamiliar to me, certainly not native prairie grasses, all of which I'd lovingly studied and knew well. Puzzled, I pressed specimens to key out over the winter.

Looking back, I realize that part of our problem was that we were thinking too much about prairie and weren't picking up what this other community — the savanna — was trying to tell us. Certainly, during these years I was worrying more about our prairie restorations than about the troublesome edges that would give rise to our savanna effort. Those prairies are another story; in fact many of our plantings in the open were going gangbusters. But I

worried, when I walked under the oaks and weeded the expanding thistle patches.

That fall we gathered more masses of prairie seed from our flourishing, open sites. We also began to gather seeds of whatever native plants were growing best in our edge plantings and in other partly shaded areas like those that were bedeviling us.

## Winter

Fortunately winter comes each year, giving me time to think. During the indoor months at the beginning of 1985, I read Floyd Swink and Gerald Wilhelm's 800-page *Plants of the Chicago Region* (1979) from cover to cover. The bulk of the book is made up of brief ecological notes for the plants in the region accompanied by lengthy lists of associated plants. In the meantime I had keyed down the three grasses that had appeared in the edge planting at Miami Woods. All three had turned out to be natives: wedge grass *(Sphenopholis intermedia)*, wood reed *(Cinna arundinacea)*, and Virginia wild rye *(Elymus virginicus)*. As I studied Swink and Wilhelm's comments on these and other miscellaneous species of our oak-edge sites, they began to stop seeming quite so miscellaneous. Many were listed as associates of each other or had third-party associates in common. I began to wonder about these species that thrived in partial shade in spite of our best efforts to grow prairie species there. And I began to wonder if they might be what we ought to be gathering and planting — at least initially — to ward off the assault of the thistles and briars.

Meanwhile, I received another prod directing my attention toward the oak groves. The previous June, John White of The Nature Conservancy's national office had sent our office a memo on critical areas recommending that one of our top priorities should be the savanna. ''Savannas are nearly exterminated, everywhere in the Midwest,'' he wrote. ''Remnants should be saved and restored.'' This was the first time, at least in my experience, that a Conservancy planner had proposed restoration. White's comments underscored the significance of what we'd been wrestling with on the North Branch.

In libraries I now found myself rummaging through old papers. What we call savanna, the early settlers often referred to as "barrens," and as early as 1863, Henry Engelmann had marveled at how rapidly these barrens grew up into trees following settlement, when the wildfires were stopped. In 1936 A. G. Vestal wondered what the plants of the barrens had been and speculated that they might still survive on wood edges, logged areas and the like.

Vestal also speculated that the savanna understory was not simply prairie, and his suspicion jibed with my experiences. When the Illinois Natural Areas Inventory looked for savanna areas worth preserving, what they sought

— and did not find — were undisturbed stands of oaks with prairie flora underneath. Maybe they were looking for the wrong thing.

If I understood what Engelmann was saying, in the absence of fire some disturbance would have been necessary for the community to maintain itself. (And it certainly was true that after a century and a half of protection from fire we do find our best savanna remnants in "disturbed" areas.) And if I understood Vestal, the key understory species to look for in these areas may in large part not be prairie species at all, but something else. Vestal mentioned American columbo *(Swertia americana)* as a possibility.

By this time I was beginning to be reminded of the experience of Dr. Robert F. Betz of Northeastern Illinois

(Above) Fire originally kept the understory free of brush and open for the savanna grasses and wildflowers. After settlement, grazing cattle kept the understory open until such areas were "preserved," as in this 1921 photo of a new forest preserve in southern Cook County. Today this area is dense brushland or young forest. In either case, the oaks don't reproduce, and the rare animals and plants of the savanna are gone.

•

(Right) Many of the original herbs of the savanna can be found today only in small numbers and in odd situations, such as along the edges of mowed horsepaths like this. For these plants, the mowing acts like a substitute for fire, and keeps back the brush.

Steve Packard

Pasture and Range Plants

# GRADUALLY I began to realize that the vegetation I was looking for did exist — as pitiful , patchy remnants at best, but it did exist.

*SWITCH GRASS*

University when he began rescuing the relic prairies of Illinois back in the 1960s. Bob had found some of the Midwest's finest ''virgin'' prairie gems in old mowed cemeteries, but at first he had met with considerable skepticism as to whether these small, degraded patches really were prairies. And indeed, where he managed to get the mowing stopped many of the sites immediately erupted with masses of Eurasian weeds. But Dr. Betz had learned to recognize the clues that indicate a vital native community surviving under the weeds, and after a few years of good management, the weeds were history and the ''virgin'' prairie looked immortal.

Betz eventually got so good at this sort of ecological rescue work that he was able to recognize surviving prairies on even more severely disturbed sites. Soon he was drawing the attention of Natural Areas Inventory ecologists to derelict land — tracts full of weedy trees, slashed by motorcycle tires, pocked with weed patches and trash mounds — as high priorities for preservation. Before Betz, before the insights gained from prairie management, most authorities had thought that Illinois black-soil prairies were gone. But now that we had learned to recognize them we were finding bits and pieces all over and were often able to bring them back to health. Is it possible that the oak openings and prairie groves could be resuscitated as well?

Encouraged by the thought, I began to construct a list of potential savanna species, beginning with those that had been appearing in our oak edge plantings. Soon I was finding some of these in the fragmentary lists in the old journals, and the old lists in turn were suggesting new species. Bebb (1860), Shimek (1911), Vestal (1936), and others provided entries. Then I considered the associates of all these as listed by Swink and Wilhelm.

I read and brooded over the 800 pages of Swink and Wilhelm like an ecological mystery novel. As my list of likely savanna species grew I sensed the possibility that a lost community was emerging from the nether edge of oblivion.

The resulting list of ''Distinctive Savanna Species,'' published informally in the proceedings of the 7th Northern Illinois Prairie Workshop, contained 122 species. Many were plants I didn't know. But as the 1985 growing season began I prowled golf course roughs, railroad rights-of-way, cemetery corners, and the edges of horse paths in forest preserves. Gradually I began to realize that the vegetation I was looking for did exist — as pitiful, patchy remnants at best, but it did exist. On nine out of ten likely-looking sites I would find nothing but rank weeds, yet in an edge or corner where I would spy a few good species, I would soon find many more. With them I would find others, often unfamiliar species, and one after another, these keyed down to the species I didn't know on that first list. I began collecting all the seeds I could find in these places.

Another insight into the savannas came in Ross Sweeney's garage that fall as we prepared our first major batch of ''savanna mix.'' Our savanna seed, like our prairie seed, was stored as gathered in hundreds of separate bags. In Ross's garage we dump this wealth of germ plasm onto big plastic sheets to make our mixes. A prairie mix — the ''dry mesic mix,'' say — ends up as a five-bushel mound of seed and fluff, which we sift and mix with rakes and shovels to prepare for planting. We'd occasionally sift the splendid golden seed through our fingers like Silas Marner, counting out the rare forbs and grasses, teaching each other how to identify them by their functional ornamentation.

Not so savanna seed. Though we now treat it differently, that first fall when we reached into the storage bags we pulled out multicolored handfuls of lumpy, oozy glop containing the red and orange fruits of tinker's weeds (*Triosteum* sp.), mushed blue-black and pink and red berries of the Solomon's seals (*Polygonatum* and *Smilacina*), purple fermenting plums and tawny hazelnuts (*Corylus americana*). There were in all more than three dozen species of fruits and nuts, dramatically making the point that we were dealing with a community that did not func-

(Left) Cutting dense brush back to bare ground at the site of a former prairie/savanna edge. The ground is then replanted to the original species of grasses and flowers. Visible here is European buckthorn. Volunteer crews of brush cutters have revealed small populations of several endangered plant species, just barely hanging on in thickets like this.

Steve Packard

(Right) After seed is gathered it is processed for planting by running it through screens to break up the seedheads. Various species need to be sandpapered, treated with acid, burned, boiled, or inoculated with specialized bacteria before they will germinate. Filling in for nature isn't simple.

Bev Modloff

tion by prairie principles, but depended, for one thing, not on wind to disperse seed, but on animals — no doubt the turkeys, deer, squirrels, doves and other species known to have been abundant on the old savannas.

These are now known mainly as forest edge species. We mixed our piles of fruits and nuts with the little sacks of rare grasses, asters, bellflowers, gentians, sedges — whatever native plants we could find in semi-shaded areas similar to those where our restoration efforts had thus far shown such dismal results. We mixed these half-and-half

with our regular prairie seed, and once again threw handfuls to the wind, back and forth through the oaks. Then waited for the next season.

## Bad Reviews

Most academics I asked to read my initial savanna paper were, to say the least, unsupportive.

One critic objected to the presence on the list of woodland milkweed (*Asclepias exaltata*). ''This is not a savanna species,'' she wrote. ''It is found in oak woods 10-30 years after grazing.'' But where was it for 10,000 years before cows arrived in the area, I wondered. The only journal editor to whom I showed the manuscript responded that my approach was ''dangerous,'' and so profoundly flawed I ought to stop working on the savanna and wait for someone who would do it right. The savanna species list should be limited to those now growing on the best surviving sites, he wrote. What's gone is gone. He was particularly concerned that someone might attempt restoration using the ''untested species'' on my list. The idea that someone thought they might be able to learn something new about a revered natural community through lowly restoration experiments seemed especially to offend these critics.

The sting of all this criticism was reduced somewhat by the fact that the list of hypothetical species was proving so useful. Now that I had some idea of what to look for, these species were precisely what I was finding in former savannas along railroad rights-of-way and the like. The list also prompted a few colleagues to do some hard thinking about old experiences. After reading the list, Wayne Schennum of the McHenry County Conservation District remembered a site he'd evaluated many years and two jobs earlier. It was a site the state Inventory had rejected as ''weedy and disturbed,'' but local people, hoping to raise money to preserve it, were seeking to build a case that it had at least some ecological significance. At the time Schennum's evaluation had been rather negative. Though the big old scattered oaks were there, the understory had neither enough woodland herbs to be a good woods nor enough prairie herbs to make it a savanna. Reviewing the new species list, though, Schennum recalled that the site may very well have been rich in the oddball species on

my list. I checked it out and found it to be a goldmine of oddballs. Since then many ecologists have come to regard what we now call the Middlefork Savanna as the finest tallgrass savanna in Illinois. It has become a top acquisition priority, and two prescribed burns have revealed a splendid mix of prairie white-fringed orchids (*Platanthera leucophaea*), woodland yellow ladyslippers (*Cypripedium calceolus*), and a preponderance of those distinctive savanna species. It was precisely that hypothetical species list that had made it possible for us to recognize this site for what it was.

Among many colleagues the early skepticism was gradually giving way to genuine interest. But one criticism was tough to shake. John Curtis, whose insightful and definitive studies of the natural communities of Wisconsin set the pattern for much of our effort to understand and preserve the Midwest's natural communities, said plainly that the savanna was a prairie with trees. He and his student, J. R. Bray, had in fact published masses of tables and statistics to make their point. They proved there was no such thing as a distinctive savanna flora — ''to four decimal places,'' as one person put it.

The authority of that proof, however, was much diminished in my judgment by the fact that the sampling had been done in unburned areas selected on the basis of some questionable assumptions, and that it was done in the 1950s, long after the real savannas were gone. But it was hard to ignore the fine botanists who had done the best it was possible to do — especially since the only contrary information was our controversial ''research by restoration'' efforts on the North Branch.

Then one day in the intellectual winter of 1985-86, following up on a lead given me by Max Hutchinson of the

Steve Packard

Handfuls of rare seeds are thrown to the wind across acres of burned ground among old oaks. Soon hundreds of thousands of rare plants will reclaim their lost ground.

Volunteers rake a prairie seed mix into the burned but still dense turf. This degraded prairie remnant is missing about half of its original species.

THE only journal editor to whom I showed the manuscript responded that my approach was "dangerous," and so profoundly flawed I ought to stop working on the savanna and wait for someone who would do it right.

Pasture and Range Plants

*BUTTERFLY MILKWEED*

Natural Land Institute, I found in an 1846 issue of an early rural journal called *The Prairie Farmer* an article by a country doctor named S. B. Mead titled "The Plants of Illinois." Without a word of prose Dr. Mead, the discoverer of Mead's milkweed, listed all the plants he had seen in west-central Illinois as he traveled his professional rounds on horseback. He began this work in 1833 as one of the region's first settlers, and most of the land he saw was unplowed and ungrazed. The list was in no particular order, and the names were incomprehensibly ancient. But scientific names can be deciphered. The important thing was that after each name was a letter or two. And a key showed that these referred to habitats: "P" prairie, "T" timber, "W" wet, "H" hills, "S" sand, and so forth. One hundred and eight entries were followed by a "B" for barrens, that central Illinois settler's name for the grassland with trees. I felt as though I'd found a Rosetta Stone for the savanna.

Once laboriously decoded, the names turned out to be marvelously familiar. They were the same sorts of plants — often even the identical species — that were growing so well in our oak-grove restorations and that appeared on my published list. Few typical prairie species had a "B" after them in Dr. Mead's "catalogue," and in every case those species had a "P" as well.

The spring of 1986 arrived on the North Branch savannas with amazing grace. Countless tiny new green cotyledons throve where only black dirt and thistles had been. The species that we had gathered and sown the previous fall, thinking they might be savanna species, and then had found on Dr. Mead's list of barrens species, were now soaking up the mottled sunlight under the old oaks. By summer of 1987 the space under and near the oaks waved with hundreds of thousands of blooming rare and uncom-

In 1846, Dr. Samuel Barnum Mead published the only known list of plant species of the original midwestern savanna. Early country doctors were botanists, because the only pharmacy at hand was local wild plants.

THE space under and near the oaks waved with hundreds of thousands of blooming rare and uncommon grasses and flowers. An oak grassland community was unfolding, almost entirely from seed we'd held in our hands

*PLAINS BEEBALM*

Pasture and Range Plants

mon grasses and flowers. An oak grassland community was unfolding, almost entirely from seed we'd held in our hands. Bottlebrush grass, silky wild rye, blue-stemmed goldenrod, tower mustard, starry campion, and big-leafed aster were among the most common species, but dozens more were also abundant. Butterflies like the banded hairstreak and great-spangled fritillary chased each other and worked the flowers where recently there had been only buckthorn, then two depressing years of bare dirt.

This was especially gratifying because many of the plants in our growing list of savanna species, though described as common in the early part of the century (Peppoon, 1927) are now listed as endangered or at least locally rare. Bev Hansen, who oversaw the savanna seed collection in 1989, often found only a few stems within a yard or two of some railroad right-of-way, horsepath, or mowed picnic grove. For most species, a handful of seed was as much as we could hope to gather in a season. But after two years of restoration, our own ground is producing vastly more seed of such refugee species than we could find elsewhere in all the thousands of acres of forest preserves.

In the drought year of 1988, many of the invading weeds shriveled pathetically, while the seeded original natives continued their advance. And in 1989, a pair of Eastern bluebirds returned here to bring up a family in their native habitat once again. They spent most of their time in the restored open woodland, picking insects, perhaps rare insects, off the lush rare grasses and flowers we'd planted. How this uncommon bird (only one pair was known to breed anywhere in Cook County the previous year) managed to find our handiwork, only nature knows. We took it as an endorsement.

Early in this century Henry Allan Gleason, one of the grandparents of ecology, wrote:

"Unfortunately for the ecologist, the prairies of Illinois were converted to corn fields long before the development of ecology and phytogeography in America, thus forever prohibiting the *satisfactory* investigation of some questions of the most absorbing interest and also of considerable importance in aiding a clear understanding of American ecology, and phytogeography in general." (1909)

A few years later he wrote:

"The only undisturbed contact of typical virgin prairie and forest observed by the writer in Illinois has been so long protected from fire that the forest margin has grown up to an almost impenetrable thicket of several species of shrubs, whose prevailingly avevectent mode of dispersal may indicate their recent arrival in the habitat. . . .

"There seems to be no authentic data on the matter, but it is entirely probable that even within the hazel margin there were numerous grasses, sufficient to feed a more destructive fire than the usual litter of leaves and dead twigs." (1913)

I think Gleason would have been happy indeed if he could have imagined what would happen half a century later — that a movement of the committed few at first, then a respected and growing element of the general culture, would begin to preserve the prairies and other rare communities he took to be all but lost. I think he would have been equally pleased to know that the efforts to restore these communities would provide a way of learning about them, providing answers to some of those questions "of the most absorbing interest and . . . importance."

In our work along the North Branch we hope that we may be helping to write a new chapter of this history through the resurrection of a complex, dynamic, splendid ecosystem that no ecologist has ever seen. ■

## Prairie Propagation Handbook • Prairie Restoration for the Beginner • How to Manage Small Prairie Fires

The people of the American Plains are giving up their lingering flatland inferiority complexes and falling in love — again — with their prairies. No damn mountains anywhere, but there are bluffs, coulees, and the terrain of the driftless areas the glaciers missed. No ocean, but enough springs, ponds and muscular rivers (not to mention the Great Lakes) to make an arid westerner think twice about the expression "green with envy." And enough ecological complexity that it is only within the last thirty years that the white man has begun to figure out just how a prairie works.

Most got turned into corn and soybeans. If more had been preserved, you wouldn't see the amount of effort being marshalled today to restore the little that remains. It's a kind of bioregional homecoming, a collective act of rehabilation — instead of wishing you were somewhere else, look at what used to be growing under your feet, and get it growing there again.

Here are three cheap, simple pamphlets to help. **Prairie Propagation Handbook** describes techniques and has a monster plant listing, alphabetically by scientific name. **Prairie Restoration for the Beginner** has a question-and-answer format, plus a listing of where the remnant prairies are, by state and Canadian province. And **How to Manage Small Prairie Fires** explains how to provide that essential ingredient — without which a prairie turns into a woodlands or a Eurasian weed patch — without losing your eyebrows or the neighbors' goodwill.
—Richard Nilsen
[Suggested by Steve Packard]

•

*Ecotypes*
Ecologists often advise the use of local seed and plant strains (closest ecotypes) to avoid possible loss or hybridizing of native ecotypes that have developed over many centuries and may have significant characteristics not yet understood. Also, there is the fear that the more vigorous southern strains may actually crowd out the local forbs and grasses, reducing the number and varieties in an abnormal manner.
—*Prairie Propagation Handbook*

•

*When is the best time to plant?*
In the northern tier of prairie states, one can plant prairie as late as June 15, and as early as September 15. Fall plantings should be made late enough in the season so that the seed will remain dormant into the winter. In the southern prairie states, seed should be in before May 15, and after October 15.
—*Prairie Restoration for the Beginner*

•

The most obvious effects of a burn are

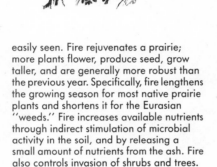

easily seen. Fire rejuvenates a prairie; more plants flower, produce seed, grow taller, and are generally more robust than the previous year. Specifically, fire lengthens the growing season for most native prairie plants and shortens it for the Eurasian "weeds." Fire increases available nutrients through indirect stimulation of microbial activity in the soil, and by releasing a small amount of nutrients from the ash. Fire also controls invasion of shrubs and trees.
—*How to Manage Small Prairie Fires*

### Prairie Propagation Handbook
Harold W. Rock, 1981; 74 pp.

### Prairie Restoration for the Beginner
Robert Ahrenhoerster and Trelen Wilson, 1988; 32 pp.

**$3.95** each ($5.05 postpaid) from Wehr Nature Center, 9701 W. College Avenue, Franklin, WI 53132

### How to Manage Small Prairie Fires
Wayne R. Pauly, 1985; 30 pp.

**$3** postpaid from Dane County Park Commission, 4318 Robertson Road, Madison, WI 53714; 608/246-3896

## Environmental Restoration

In 1988, John Berger and his organization Restoring the Earth sponsored a landmark conference in Berkeley, California, that helped put environmental restoration on the map. This book, a collection of the technical papers from the conference, demonstrates the amazing diversity of restoration projects, and of the people and groups doing them. Restoration ecology is giving amateur and professional earth workers a common place to hang their hats.
—Richard Nilsen

### Environmental Restoration
John J. Berger, Editor
1990; 298 pp.

**$19.95** ($21.95 postpaid) from Island Press, P. O. Box 7, Covelo, CA 95428; 800/828-1302 (or Whole Earth Access)

•

Woodland restorations vary in composition and structure depending on their purposes. . . . Pest species, in this context, are plants that interfere with such restoration goals. The plants may be native or nonnative, and the same species may operate as a pest in one situation but not in another. . . . These effects are reinforced by microclimate changes brought about by the presence of pest species. In woodlands, such changes can include increased shade, reduced moisture availability, and changes in soil fertility. For example, in the Midwest, Eurasian species often have a longer growing season than American species. Woodland spring ephemerals require high sun levels to bloom and produce food. Such levels typically occur under the leafless April and May canopies of native trees and shrubs. These herbs are weakened when their habitat is invaded by Eurasian species which produce solid canopies as early as April in parts of the Midwest.

**Harvesting a wild seed assortment with a Grin Reaper.**

# EARLY AMERICAN TREEPLANTING
## From Apple Seeds To Arbor Day

*"First of all, Indian and European cultures were in direct conflict. One depended upon the woods or plains; the other upon plowed fields and pastures. One exploited basic natural resources for livelihood; the other relied upon cultivated crops. . . . The only physical things, in fact, which survived many of the settlers were their cleared and eroded fields."*
—*Thomas D. Clark*, Frontier America *(1959)*

## BY RICHARD SASSAMAN

*Before restoration, there was reclamation, rehabilitation, soil stabilization and reforestation; terms not synonymous, but all containing the idea of repair or improvement. There was also good old-fashioned treeplanting. While most 19th-century Americans were busy felling trees, a few were planting seeds that would bloom in soils and minds into the next century and beyond.*

*Richard Sassaman lives in Bar Harbor, Maine.*
*—Richard Nilsen*

European immigrants to 18th-century America regarded the great virgin forests they found as "wild"erness that had to be tamed. Trees took up valuable space needed for fields of growing crops or grazing livestock, and could be turned into firewood, log cabins, split-rail fences, or crude forts used for protection against the natives.

This attitude figures prominently in perhaps the first great American myth, which describes how young George Washington killed his father's cherry tree. Invented by Mason ("Parson") Weems in 1806 (after Washington's death) the apparently simple fiction can be interpreted on several levels. "I cannot tell a lie," George says, running to his father. He *wants* to tell the truth, to chop away at his "family tree" and openly display rebelliousness. (In the original version, George takes his axe to an "English" cherry tree.)

The legend also depicts the "Father of his Country" on his birthday, a convenient symbol for the newborn United States. Washington, given a nice new axe as a present, must immediately make use of it, just as Americans — given both the technology and the opportunity — felt they had a "manifest destiny" to civilize the wilderness by cutting down the forests.

Under these circumstances, few Americans concerned themselves with problems of deforestation. In the early 1680s, William Penn did require that an acre of forest be maintained for every five acres cleared on lands he granted, but no records show that this provision was ever enforced.

Another rare conservation incident occurred in 1799 only for national security reasons, after President of the Continental Congress John Jay warned that Navy ship timbers and masts might become unavailable in the near future unless suitable tall trees remained protected. Congress authorized President John Adams to spend $200,000 buying live-oak reserves along the South Carolina and Georgia coasts, probably the first Federal money spent for purchasing timberland.

Not until 50 years later would the U.S. *Patent Office*, oddly enough, issue a report dealing with the impact of accelerating forest destruction and its attendant effect on water flows. The report noted that timber waste in the nation ''will hardly begin to be appreciated until our population reaches 50,000,000. Then the folly and short-sightedness of this age will meet with a degree of censure and reproach not pleasant to contemplate.''

The Civil War also provided an impetus for forest protection, when far-flung military operations both required a great deal of wood and destroyed much valuable timberland.

Because of these and other warnings — most notably a paper written by Frederick Starr in 1865 for the Department of Agriculture — the importance of tree planting finally caught the public's interest. Starr, a preacher, couched sections of his work in apocalyptic language, warning that ''the nation has slept because the gnawing of want has not awakened her. She has had plenty and to spare, but within thirty years she will be conscious that not only individual want is present, but that it comes to each from permanent national famine of wood.''

> ''Of Jonathan Chapman
> Two things are known,
> That he loved apples,
> That he walked Alone. . . . ''
>
> —Rosemary and Steven Vincent Benet,
> *A Book of Americans* (1933)

One early American who never doubted the urgent need for tree planting in this country, John Chapman of Leominster, Massachusetts, became widely known during his lifetime as ''Johnny Appleseed.'' Although he's remembered today mostly as a mythical character of legend, Chapman was real, not an advertising creation like Paul Bunyan or the Jolly Green Giant. He stood 5' 9'' tall, and had blue eyes.

Born in 1774 (almost simultaneously with the nation), Chapman soon headed west from New England. (One of 12 surviving children from his father's two marriages, he might have been looking for peace and quiet.)

Readers seeking the truth about Chapman are indebted to biographer Robert Price, who spent 25 years on the pioneer pomologist's trail separating fact from fiction before writing his *Johnny Appleseed: Man and Myth* (Indiana University Press, 1954).

According to Price, ''In spite of all the delightful imaginings by several generations of Johnny Appleseed cultists . . . not a single fact is known today, based on either record or direct tradition, that will bridge the wide gap between [his] certifiable origins in [Massachusetts] and his probable appearance in the upper Allegheny Country of northwestern Pennsylvania in 1797.''

John lived in this sparsely populated land for about a decade. Then, deciding his neighborhood had become too crowded, he moved on to frontier Ohio, where in addition to attracting attention for growing apple trees he became a hero during the War of 1812, warning settlers against Indian attacks.

Ultimately Ohio became overrun with settlers as well (its population rose from 45,365 in 1800 to 937,903 in 1830), so Chapman continued on to northern Indiana. He died at a friend's cabin near Fort Wayne in March 1845.

Curiously, the first mention in print of Johnny Appleseed appeared in England, and for religious reasons. Somehow he had become an ardent disciple of the Swedish religious philosopher Emanuel Swedenborg (at a time when fewer than 400 Swedenborgians lived in the U.S.), and often corresponded with or visited church officials.

The article, published in 1817 in a Swedenborgian Society report, began:

''There is in the western country a very extraordinary missionary of the New Jerusalem. A man has appeared who seems to be almost independent of corporal wants and sufferings. He goes barefooted, can sleep anywhere,

**The only existing drawing of John Chapman, 1850.**

in house or out of house, and live upon the coarsest and most scanty fare. He has actually thawed the ice with his feet.''

Chapman on his wanderings presented a picture of true poverty, wearing ragged old clothes while going from place to place, accepting a meal here or a spot on the floor by the door there.

Despite this myth of the wandering gypsy, however — and it is true that Chapman rarely bothered with possessions — Robert Price followed the historical Johnny Appleseed through a paper trail of legal documents (land deeds, loans paid, etc.), as well as frontier diaries and other writings. Adding up statistics, he notes that ''Altogether [Johnny] owned either by deed outright or on long-time leases no less than twenty-two properties totaling nearly twelve hundred acres.''

Johnny Appleseed is usually remembered today as an eccentric old man; one rumor had it that he'd been kicked in the head once by a horse, which didn't help his mental stability. ''He conforms to none of the national stereotypes and illustrates nobody's theories,'' Edward Hoagland wrote to explain this common misunderstanding.

Also, Americans today probably don't understand how much the early settlers appreciated apples, which in Price's

words are now ''relegated in normal modern diets chiefly to side dishes and casual eating.'' Apple butter would keep for winter use, apple brandy was a cash export sent downriver to New Orleans, apple cider a social favorite, and apple vinegar the basic pioneer preservative.

But in the country of the blind, the one-eyed man is king. And back then, for almost 50 years in the country of the apple-hungry pioneers, John Chapman lived like a king.

''A part of the health of a farm is the farmer's wish to remain there. His long-term good intention toward the place is signified by the presence of trees. A family is married to a farm more by their planting and protecting of trees than by their memories or their knowledge. . . .''

—Wendell Berry,
*The Unsettling of America* (1977)

By the time early settlers reached the prairies of Illinois, then continued on to the Great Plains, their attitude toward trees had changed. Back East the majestic forests may have been in the way, but Nebraska and Kansas sodbusters longed for houses made of wood, and for trees that would provide fuel, give shade, help control soil erosion, and serve as windbreaks.

One early Midwesterner who worked hard to overcome

*(Above) Sixth-grader Richard Sassaman plants an oak tree at his school on Arbor Day, 1961.*

*(Right) Richard and the tree today.*

# Me And  My Tree

The Chinese proverb says a person must do three things to lead a full life: plant a tree, write a book, and have a child. I achieved the first of these goals on Arbor Day in April 1961, when as a sixth-grader I was chosen to be the ''designated planter'' for my school in Harrisburg, PA.

Back then it didn't occur to me to ask why only one student was planting a tree, while all the others merely watched. I stood happily in front of news cameras with the principal and the P.T.A. president, throwing a shovelful of ceremonial dirt over the roots of a brand-new oak tree (the symbol of the P.T.A.).

That afternoon I also recited a three-paragraph speech, a copy of which my mother never threw away, which someone (maybe her) must have helped me write. Both a remarkable historical document, and also an example of the great gibberish people will spout when urged to do so by ghostwriters, my speech went like this:

''I am happy to accept this tree on behalf of the children of Northside School. We are proud of our parents and teachers for having made this occasion possible through their interest in us.

''My classmates and I have taken our roots of learning from this school, and this regional lack of trees, Julius Sterling Morton, actually was born in New York State in 1832. Two years later his family headed west to Monroe, Michigan, and then to Detroit.

"Owing to his independence of the constituted authorities," as the *Dictionary of American Biography* puts it, Morton was expelled in his senior year from the University of Michigan. (He later received an A.B. degree retroactively.) After marrying at age 22, he celebrated his honeymoon by looking for a new home in the Nebraska Territory.

As editor of the *Nebraska City News*, the area's first newspaper, Morton continued his independence of authority. Inspired by a local group of expatriate Germans who had emigrated from the treeless steppes of Russia, where the first "shelterbelt" tree rows had been cultivated, the editor began tirelessly promoting tree planting among his fellow Nebraskans.

"There is beauty in a well-ordered orchard which is a joy forever," Morton once said in an address to the State Board of Agriculture. Adding that "Orchards are missionaries of culture and refinement," he said, "If I had the power I would compel every man in the state who had a home of his own to plant and cultivate fruit trees."

Perhaps seeking that power, Morton ran for public office numerous times: territorial legislature, U.S. Congressman, U.S. Senator, state governor. Sometimes he won, sometimes he lost. Appointed secretary of the Nebraska Territory by President James Buchanan in 1858, he served until 1861, never giving up his interest in forestry.

In January 1872, Morton presented his "Arbor Day" resolution to the State Board of Agriculture. "Resolved, that Wednesday, April 10, 1872, be . . . especially set apart and consecrated for tree planting."

"To urge upon the people of the State the vital importance of tree planting," the resolution offered a prize of $100 to the county agricultural society that planted the most trees, "and a farm library of twenty-five dollars' worth of books to that person who, on that day, shall plant properly, in Nebraska, the greatest number of trees."

Morton named Arbor Day after the Latin word for "tree." Some members of the Agriculture Board thought his event should be called "Sylvan Day," but the resolution passed as written. And on the first Arbor Day, a great success, over a million trees were planted by Nebraskans, unfortunately not including J. Sterling Morton.

The founder, who personally planted thousands of trees during his lifetime, had ordered 800 saplings specially for

like the oak planted here today, we hope to branch out and become sturdier, stronger, and more useful in the future.

"Years from now, it is our hope to find a full-grown tree giving shade, beauty, and symbolic strength to our school. Thank you very much!

Despite the hogwash, and the fact that I'm not sure how sturdy and useful I've been myself over the past 28 years, the tree — now about 40 feet tall — really *has* branched out to give shade and beauty to the school. And despite the fact that only God can make a tree, and that I didn't even dirty my hands planting it, I've grown to think of that oak as "mine."

Whenever I return to central Pennsylvania I go by to see my tree; during the breakdown at Three Mile Island a dozen miles away, I sat in Seattle and worried about it.

Recently I stopped by Northside again and visited with current principle Ken Beard, who's been there since 1966. We looked at old photos of people planting trees, and he told me the school P.T.A., for at least the last 15 years, has planted a tree to honor every retiring teacher. Tree planting is an honorable enterprise, one that you'll get even more enjoyment from as the years go by. ■

*Richard Sassaman*

the initial holiday ceremonies, but these got lost temporarily in transit, and didn't arrive. ''They will come soon and then I will put them out,'' he said resignedly.

Morton spent the first Arbor Day writing to a newspaper in Omaha, saying, ''There is a true triumph in the unswerving integrity and genuine democracy of trees. . . . Trees planted by the poor man grow just as grandly and beautifully as those planted by the opulent. . . . The wealthiest and most powerful potentate on earth cannot hire one to speed its growth or bear fruit before its time.''

Within a few years Arbor Day had become popular in Kansas, Tennessee, Minnesota, and Ohio as well. During a remarkable first ceremony held in Cincinnati in 1882, almost 20,000 young students assisted in planting a grove of trees, with each tree being named for a distinguished person. This event more than any other established the custom of Arbor Day being observed by schoolchildren.

Largely because of his work with Arbor Day, Morton was named the third U.S. Secretary of Agriculture in 1893, serving under President Grover Cleveland. He died in April 1902 at the home of his wealthy son Paul, who founded the Morton Salt Company.

After J. Sterling Morton's death, his stately home ''Arbor Lodge'' in Nebraska City, on 65 acres planted with over 160 species of trees, became a Nebraska state park.

Today every state honors Arbor Day during a time best suited for planting trees, as do 15 countries including Canada, China, France, Great Britain, Israel, Japan, and the Soviet Union. In 1885 Nebraska selected Morton's birthday, April 22nd, as its Arbor Day; California chose March 7th, the birthday of Luther Burbank. National Arbor Day is observed each year on the last Friday in April.

In his *A Brief History of Forestry* (Toronto, 1911), Bernhard Fernow complained that ''Arbor days have perhaps . . . had a retarding influence upon the practical forestry movement, in leading people into the misconception that forestry consists in tree planting, in diverting attention

**Julius Sterling Morton founded Arbor Day — now observed in all 50 states and 15 foreign countries.**

from the economic question of the proper use of existing forest areas, [and] in bringing into the discussion poetry and emotions, which have clouded the hard-headed practical issues, and delayed the earnest attention of practical business men.''

These days the holiday, overshadowed in recent years by more glamorous events like Earth Day, seems more practical than ever. Thanks to widespread drought, rain forest destruction, the greenhouse effect, and global warming, tree planting has once again become a hot subject.

As J. Sterling Morton explained over 100 years ago, ''Other holidays repose upon the past. Arbor Day proposes for the future.'' ■

# Global Warming/Global Warning: Plant the Right Tree

*by Marylee Guinon*

IN AN EFFORT to reduce greenhouse gases and compensate for global warming, it is likely that billions of seedling trees will be planted during the next few years. The dedicated people doing the planting hope the trees will thrive into maturity and act as carbon sponges to fix atmospheric carbon dioxide. Often non-native species are planted. These can naturalize and spread as escaped exotics, displacing native species and even driving some natives into extinction. But even when the correct native species are planted, the local biodiversity can still be lost or destroyed if no one is paying attention to the genetic source of the seedling tree stock.

Genetics conservation is the protection and preservation of the genetic raw materials of adaptation and evolution that species and ecosystems depend upon for long-term survival. Unlike humans, plants cannot modify or move from their environments as they contend with hardships like droughts, freezes, and even ice ages. Plants are able to survive on specific sites because they accumulate the large amounts of genes their ancestors evolved over millennia for life in those places. Many individual tree species are known to contain *several times* the genetic variation of our own human species. These huge pools of genes are like a bag of survival tricks, that can be called upon if needed during lifetimes that often last centuries.

Tree planters need to be aware that within a given species, each stand of trees contains a unique diversity of genes appropriate to that site, and that it is only this pool of genes that is likely to survive there through future hardships — such as global warming — over the lifetime of the trees and the lifetime of the species. Tree planting that ignores genetics conservation hurts two ways — it can lessen the variations that exist within trees on a site, and it can contaminate a local gene pool with ill-adapted introductions from somewhere else. The differences are hidden away in the genes; non-local seedlings and their offspring are indistinguishable to the eye from the local inhabitants.

In extreme cases, tens of thousands of seeds are collected from just a few parent trees. This can seriously reduce genetic variation and result in inbreeding and widespread

mortality in the future. Conscientious tree planting requires more start-up preparation and expense, but the benefits will be realized over the long term, perhaps centuries.

Before undertaking a large-scale planting, it is worth preparing genetic conservation guidelines to insure that the planting stock is compatible with ecology and genetically appropriate to the site. A few private consultants can prepare these guidelines for public or private projects. Your local cooperative extension office may be able to refer you to plant geneticists who do this work, or to agencies familiar with such guidelines. ∎

*Ms. Guinon is a private consultant specializing in genetic conservation. She can be contacted by writing to Sycamore Associates, 910 Mountain View Drive, Lafayette, CA 94549.*

*—Richard Nilsen*

## The Simple Act of Planting a Tree

*This is as much a book about organizational and motivational skills and fundraising as it is about tree planting. The subject is urban forestry, and the folks at TreePeople in Los Angeles have been at it since the early seventies. Their most notable achievement was planting a million trees in LA in time for the '84 Olympic Games. Trees in cities don't make it without people to protect them, so urban forests are really a living embodiment of community spirit. This book replaces the earlier **Planter's Guide to the Urban Forest**, and is filled with those niggly details most people overlook when they write how-to books. The degree of organizational transparency documented here makes it worthwhile for anyone involved with a volunteer group.*
*—Richard Nilsen*

•

The best time to plant a tree was twenty years ago. The second best time is now.
*—Anonymous*

•

*Environmental restoration*
It doesn't matter how you do it, but creating your vision can be the most powerful part of your entire project, and it deserves a lot of respect. If you've never taken anything past the good-idea stage, try this.

Think about the project and envision it as complete. Examine all the elements as they would be if the project were finished. Spend some time savoring those images. Now work backwards. Build a bridge back to the present. From the point of completion back, at every phase, see if you can see who is doing what and what resources or tools are being used. Without limiting yourself by wondering how you can afford it,

**The Simple Act
of Planting a Tree**
TreePeople with
Andy and Katie Lipkis
1990; 240 pp.

**$12.95** ($14.70 postpaid) from
Jeremy Tarcher, 5858 Wilshire Blvd./
Suite 200, Los Angeles, CA 90036
(or Whole Earth Access)

list the kinds of people and resources you envision you'll need. As the vision unfolds, ask yourself, "How might we do that?" Limiting thoughts will come up, but don't let them stand in the way of this process.

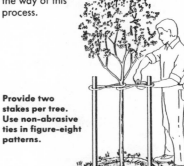

**Provide two
stakes per tree.
Use non-abrasive
ties in figure-eight
patterns.**

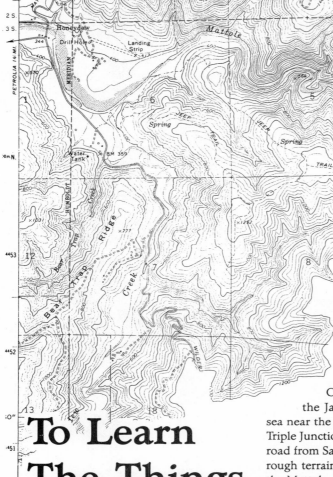

# To Learn The Things We Need To Know

## Engaging The Particulars Of The Planet's Recovery

*BY FREEMAN HOUSE*

**T**HE MATTOLE RIVER runs coastwise south to north for 64 miles, through a wrinkle in the North American crust formed as the Pacific plate collides with it and dives, pushing up the King Range. Just to the north is Cape Mendocino, California's westernmost point, where the Japanese and California ocean currents meet. Under the sea near the mouth of the Mattole, three fault lines meet to form the Triple Junction, the most seismically active spot in the state. The coast road from San Francisco to Portland takes a major detour around this rough terrain; it's an hour from any major highway to most parts of the Mattole. Redwoods grow in its fog-washed headwaters, and a rich mixture of Douglas fir and hardwoods elsewhere. Of the 2,000 or so people living here now, some two-thirds of them have migrated here in the last twenty years, as large sheep and cattle ranches have been subdivided into homesteads.

In the late seventies, a few people began to observe that the native Mattole king salmon population was diminishing in an alarming way. In recent memory it had been the local custom to gather the few large fish it took to make a winter's supply from a migration that arrived each November and December in seemingly limitless numbers. Now only the hardiest of outlaws was gaining occasional protein this way.

The Mattole run was one of the last purely native "races" of salmon in California, largely because the river was so remote that the state Department of Fish & Game (DFG) had never gotten around to stocking it with hatchery fish. In valleys where salmon run they play a large and dramatic part in the spectacle of life. Their value as food only

---

*Our ancestors never did get good at living in harmony with the land. With such an immense continent to settle, they never had to. Today, warning lights flash at almost every turn and that brash "I'll do it myself" pioneering gumption is more of a hindrance than a help; yet it persists as a vital part of who we are. This article describes a rural community busy redirecting that Daniel Boone spirit to make living on and from the land feasible into the 21st Century. Here restoration is the essential first step in a concurrent process of social and economic transformation.*

*Bioregionalist Freeman House lives in Petrolia, California, where he is co-founder with David Simpson of the Mattole Watershed Salmon Support Group and an employee of the Mattole Restoration Council.*
*—Richard Nilsen*

Janet Morrison, chairwoman of the Mattole Restoration Council, surveys late winter floodwaters on the Mattole River in 1985. Bank failure threatens the county road behind her, the main artery in the valley. With funding from Humboldt County and the California Coastal Conservancy, the Restoration Council riprapped 375 feet of this bank the following summer, using a row of five-ton boulders for a base.

Randall Stemler

begins to justify the depth of feeling that some people have for them. People who wouldn't dream of eating them get moon-eyed at spawning time, and will sit quietly in a drizzle for longer than is reasonable in order to see one jump. It became vitally important to some people in the Mattole valley to attempt to reverse the decline. It was important in terms of maintaining this most visible celebration of the mosaic of wild life in the valley, and it was important to maintain this remnant of genetic diversity for the health of all Pacific salmon. The response was to take it on, to attempt to puzzle it through, to learn whatever needed to be learned to make the king salmon population viable once more.

The few people who undertook this challenge made up for their small numbers with a large name — the Mattole Watershed Salmon Support Group. They were flying in the face of the common wisdom of the time, which was laced with despair. In the two decades between 1950 and 1970, something more than three-quarters of the Douglas fir and redwood trees which held the watershed's slopes in place had been cut for timber. Enormous amounts of bare soil had been exposed to disastrous amounts of rainfall. Starting in the flood year of 1955, hillsides began to slide into the river system, a process which continues as much as twenty-five years after the original logging disturbance, as root wads rot and cease to lend their tensile strength to steep slopes. Deep pools and channels in the river had filled up with silt; the river jumped its banks,

taking out whole stands of riparian growth which had shaded and cooled the water.

The clean gravels that salmon require for spawning and the deep pools the young fish needed to grow in were gone, and anyone with eyes for it could see the destruction in the new broad and cobbled floodplains in the lower stretches of the river where farmland had been. This was the source of the despair; the processes that had been cut loose by the too-rapid deforestation of the basin were apparently too huge to be engaged by humans with fragile limbs and frugal means. Most people were willing to ride out their assumption that nothing much could be done. If we were willing to make fools of ourselves, we would be given the opportunity, but not much else.

The salmon group worked from the assumption that no one was better positioned to take on the challenge than the people who inhabited the place. Who else had the place-specific information that the locals had? Who else could ever be expected to care enough to work the sporadic hours at odd times of the night and day for little or no pay?

Working symptomatically, we discovered a low-tech decentralized tool in the streamside salmon incubator used previously in British Columbia and Alaska, which treated the problem of silted-in gravels by imitating the ideal

The first thing we learned from salmon was the importance of the watershed as a unit of perception.

Wildlife of the Rivers

personnel; and even more difficult for some (though not all) agency-employed biologists to accept the possibility that non-credentialed locals might be able to handle some of the field work that had been their province only. There was one point in our negotiations when one of the more stiff-necked wildlife managers was insisting that we needed a fisheries biologist to drive any truck transporting live fish. Our quest for legitimacy finally took us to the office of California's then-Secretary of Resources, Huey Johnson. Huey took a chance on us. A year and a half after the salmon group had first approached the state, we were ready to put away our briefcases and put on chestwaders.

When they leave the Mattole for the open sea in the late spring or early summer of their first year, it takes as many as 200 king salmon to make up a pound of salmon flesh. When they return as adults after spending three to five years in the Pacific, one fish often exceeds twenty pounds and an occasional one may approach forty pounds of muscle, will, and utterly exotic intelligence. We had arrived at an agreement with the state which allowed us to capture twenty female salmon, each one carrying three to five thousand eggs, and enough males to fertilize them. The spawners run upriver in mid-winter, preferring the obscurity of muddied, rising water, the obscurity of night. A trap and weir close off the river, but only temporarily. Once the waters rise to a certain point, the structures must be moved out of the channel or they will be washed away. Once trapped, the salmon must be moved to safety and held until the eggs are ripe and ready for fertilization.

To enter the river and attempt to bring this strong creature out of its own medium alive and uninjured is an opportunity to experience a momentary parity between human and salmon, mediated by slippery rocks and swift currents. Vivid experiences between species can put a crack in the resilient veneer of the perception of human dominance over other creatures. Information then begins to flow in both directions, and we gain the ability to learn: from salmon, from the landscape itself.

The first thing we learned from salmon was the importance of the watershed as a unit of perception. If salmon organize themselves so clearly by watershed, wouldn't it make sense for us to organize our efforts similarly? Salmon are not only creatures of unique watersheds, adapted so that generation after generation responds to the timing and flow of utterly specific rivers and tributaries — but they are also dependent on watershed processes in general. During their reproductive time in fresh water, salmon live at the top of the aquatic food chain, but at the bottom, so to speak, of fluvial and geological processes. The success of incubation depends on the availability of river bottom gravels free of fine sediments. The survival of juveniles depends on the presence of cold deep pools cut down to bedrock.

In the Pacific Coast Range, new mountains are still rising out of the ocean bottom at the rate of two to four meters per millenium, and the soft sea silts have rarely had time to metamorphose into competent rock which might stay in place against the winter storms that wash most of the

natural situation. These incubator systems — or hatchboxes, as we were quick to call them — fed filtered water from the client creek through select clean gravels in a box the size of a pickup truck toolbox located by a creek in a neighbor's yard. Cheap to build, without moving parts or external sources of power required, the hatchboxes proved to be relatively trouble-free. They could accommodate as many as 30,000 fertilized eggs and consistently deliver a better than 80-percent egg-to-fry survival rate — compared to less than 15 percent survival in the mud-stricken river.

If we were to maintain the native adaptations of the populations we were hoping to enhance, we would have to take our eggs from native stock in the wild, rather than accept eggs from another watershed or from homogenized hatchery stock. It was around this idea that we first began to encounter official resistance to the notion of locals and non-professionals dealing directly with nature. It was difficult for the DFG to break precedent and allow non-licensed civilians to put a net or trap in fresh water, a procedure heretofore strictly forbidden to non-agency uplift back to sea. A ten-year storm, combined with any one of the frequent earthquakes which inform this coast, can cause a landslide which will change the course of the river and alter the pattern of salmon reproduction for several human generations. Combine these conditions of the fundament with a ranching technology that requires a few hundred feet of dirt road for every head of stock; with a timber economy that makes it cheapest to build a road to every tree and remove all vegetation from the slopes; with a homestead ethic that can rationalize miles of benchcut road to protect the privacy of each and every American home, and you have a recipe for the kind of catastrophic impacts we were observing.

To nurture the health and natural provision of the wild salmon population, the salmon were telling us, we were going to have to understand them as an integral part of their habitat, and that habitat was the entire watershed, extending all the way to the ridgelines above us, including the human settlements. In order to address the aquatic habitat, we would need to keep the topsoils on the slopes where they could grow forests and rangeland ecosystems, meat and vegetables, and out of the streams, where they killed fish. We would need to attempt to reduce the amount of silt entering the riverine system each year to below the amount which winter flows were able to flush out. Above all, as humans, we needed to learn to take our meat and wood in ways that didn't cancel out the potential for natural provision and other relationships

Biologist Gary Peterson milks milt from a male king salmon into a bucket containing roe from a single female. To insure against possible infertility and to provide optimum genetic diversity, milt from several males is used for each female. Water is then added to the eggs and they are placed in an incubation box.

Mattole Salmon Group

The author inside a homemade salmon trap in the Mattole River in December. Spawning salmon migrating upstream are guided into the trap by a weir that fences off the river. The Salmon Support Group traps only about five percent of each year's run. The screen on top of the trap keeps salmon from jumping out, and prevents otters, herons and raccoons from getting in.

Judi Quick

within our own habitat. Finally, the salmon were telling us, what was good for them was good for us. Both species benefit from healthy watersheds and an extended sense of commonality. "Ladies and gentlemen," they were saying, "please, let's get serious about this business of coevolution."

Bioregionally, there are numerous ways to define "your" part of the planet — physiographic areas, the ranges of species of plants and animals, climatic zones, human language groups. Ecological responsibilities, in individual or social terms, can be most successfully undertaken in the context of specific places. The human organism demands, finally, that its intuitions and ideas become embodied through physical perception, and the landscape surrounding generally offers us the best proof of our immersion in biospheric life.

In this case, the range and requirements of a particular race of salmon had defined a context for our efforts, and the 306-square-mile watershed of the Mattole offered a scale in the range of the possible. If the experience of the whole place remained beyond the perception and understanding of any one person, the drainage divided itself into more than sixty tributaries, many of which were home to groups or individuals interested in the demands of stewardship, and interested too in the possibilities of an identity extending beyond parcel boundaries.

The attempt to engage ourselves with a salmon run shifted almost at once from a symptomatic, technical-mechanical approach to a systemic, multi-leveled, ecological ap-

proach. Focusing on the crisis of another species had boomeranged into the need to take a close look at our own social organizations and economic activities. This latter was no new insight, of course. The same conclusions had been reached by everyone from the United Nations to Earth First!. The difference was the conviction that through engaging the fundamental processes of a particular place, we might discover the appropriate models for our own activities and organization. Adopting the conceit that our restoration and enhancement projects would hasten the process of watershed recovery, those same projects might be the very means by which we learned what we needed to know in order to live integrated lives in living places.

What easy rhetoric! What facile promises! The long list of skills required for the task quickly turned our leap of consciousness into a scramble for data. But to my surprise and pleasure, much of the expertise we thought we lacked was found among the 2,000-plus people already living in the watershed. Some of my fondest memories are of whole days spent once a month by twelve to twenty people, training ourselves for the initial job of inventorying salmonid habitat.

A consulting biologist living on Mattole Canyon Creek gave us lectures on Odumesque energy budgets that were full of sunlight, nearly understandable, and wholly inspirational. A self-trained naturalist gave us descriptions of mycorrhizal relationships so rhapsodic that some of us fell in love. When the information we needed wasn't

Riprapping riverbank near the headwaters of the Mattole. This is one of 17 sites, all visible from the county road, where stabilization was done by member groups of the restoration council. All rock was positioned by hand, in a kind of large-scale drywall masonry, and has become a source of pride for local residents.

Through engaging the
fundamental processes of
a particular place, we
might discover the
appropriate
models for our
own activities and
organization.

The Wilder Shore

available in the neighborhood, we never had to go far to find a helpful technician: biologists from the DFG, a geomorphologist from the Forest Service, and most especially geologists and hydrologists from Redwood National Park, that planet-class laboratory in landscape rehabilitation only a hundred miles away. The agencies proved rich in this kind of talent, and were sometimes cooperative to the point of dipping into small public-relations budgets to send their people our way during working hours. Often, too, the scientists showed up on weekends, anxious to share and test their ideas in the field.

Most of the skills we needed were gained more by experience than training or education. For all the headiness of ecological relationships and hydrological theories, for all the stretched mental landscapes of geological time, when it comes right down to engaging watershed recovery processes, you'll most often find yourself with a shovel in your hand or in conversation with a backhoe operator. The great days of the worker-owned tree-planting co-ops were past, diminished by hostile legislation, but some of the workers were still around, and no one is better fitted to organize a large-scale treeplanting or take on the risks and intricacies of a government planting contract.

As the physical effort grew, with crews engaging in salmon enhancement, habitat repair, erosion control, and reforestation, the need arose for a new sort of organization based on watershed priorities. On one level it had to serve the interests of various factions among the human community interested in engaging recovery, including groups organized around specific tributaries, groups organized around jobs, land trusts, and schools, plus civic groups devoted only peripherally to land-based issues. The larger need, however, was to invent a process for developing a

shared perception of the real ecological parameters of our riverine watershed, to make long-range plans, to make consensual decisions about projects and, increasingly, to take positions on complex issues.

Which should be approached first — the accumulated sediments choking the estuary, or the new sediments being introduced upstream?

(Answer so far: both.)

If habitat recovery can't be monitored, how many additional fish should we be introducing?

(As many as we can without violating our general guideline of closely imitating natural processes.)

Knowing that less than 10 percent of the original forest complex remains uncut, should we risk assuming a high moral ground we have no way of implementing by calling for a moratorium on the logging of ancient forests, and thereby risk alienating the largest private landowners in the watershed?

(Yes.)

The Mattole Restoration Council was formed to serve these ends, a consensual council with thirteen member groups at this writing. Monthly meetings rarely manage to remain serene deliberations on the ecological parameters of the basin, however. The Council has, over the years, arrived at a position which recognizes the need for localized ecological reserves, based on the perception that genetic diversity is more site-specific than is generally assumed, and on the exact knowledge of how little of the original forest and range complex is left. At the same time, the Council embraces a vision of the most desirable future, wherein productive lands stay in production; forests continuing to produce timber and fisheries, rangeland continuing to provide forage, and stabilized agricultural lands producing farm products.

The need for reserves outrages industrial timberland owners, and the desire for an economically productive landbase over whch local residents maintain some control makes environmentalists jumpy. Often the Council finds itself stretched thin; monitoring and resisting industries' and government agencies' desire to harvest every mature tree, at the same time as it is participating in a statewide attempt to invent a set of sustainable and restorative forest practices. Either one of these processes will eat up a lot of volunteer time, and they present themselves over and above the Council's stated goals of supporting a certain level of active rehabilitation work, of attempting to meet contract deadlines, and of developing and distributing the baseline information we need if we are to engage our watershed in a manner which restores it to previous levels of health and productivity.

The process of researching the biotic and geological realities of natural places is likely to surprise anyone undertaking it for the first time. Our own experience has led us to describe watersheds and other natural areas as unclaimed territories, so sketchily are they documented. When the DFG first opened its files to us in 1980, we were disappointed to find how scanty the salmonid habitat data for the Mattole River was, but not surprised once we came to the realization that two counties with a combined area the size of Delaware were being monitored by a single biologist. If we were to approach the restoration of native salmon populations in any sort of systematic manner, we would have to do it without historical baseline data, and in order to understand the current situation we would have to generate the data ourselves. Later, when we needed to understand just how much old-growth forest remained in the watershed, we encountered a similar situation. Antagonistic theories and counter-theories about how much ancient forest was left were getting a lot of press space, but when we took a closer look, *no one knew what was out there.* (A little later, the U.S. Forest Service found itself with the same problem.)

In both these cases, we were able to find some support for finding out, and we were faced with the choice of hiring professionals to do the surveys, or training local residents. Either option would cost about the same, the added local knowledge balancing out the cost of training amateurs in survey techniques. We went with the locals, reasoning that as a watershed population we would have gained an array of skills that had a value beyond the data we would collect.

Even when the information you need is available, it will almost always be in the wrong context if you are filling in the map of a natural area. No wonder people find themselves alienated from planetary processes, when all the information for their part of the planet is filed by township and range, rather than river and mountain. The Mattole watershed, for instance, extends into two counties, two jurisdictions of the DFG, and overlaps with about half of a Bureau of Land Management holding. Research

the distribution of a species and you'll find the data laid out for the state of California, in an amalgam of ranges and townships, rather than in an intertwining complex of habitats. Try to find out who-all lives in the Honeydew Creek drainage. You'll find the owners listed alphabetically, rather than by where they live. And so on.

By spending the time to reorganize biotic, geologic, and demographic information into a watershed context, we are ritually reanimating a real place that had become totally abstracted. Our maps of salmonid habitat, of old-growth distribution, of timber harvest history and erosion sites, of rehabilitation work, our creek addresses for watershed residents; all these become, when distributed by mail to all the inhabitants, the self-expression of a living place.

The hatchbox program is entering its tenth year now, and it has not been the quick fix our early naivete allowed us to hope for. The salmon population turned out to be even more diminished than we had guessed. If we aren't yet able to put a number on the volume of Mattole topsoil that is delivered to the Pacific each year, we do know where it's coming from, and where in the river it is tending to settle out, jamming up biological and hydrological processes. Thousands of trees have been planted, thousands of tons of rock moved to armor gullies and streambanks. A whole generation of elementary-school kids has released a lot of salmon into the wild as part of each of their eight school years. Some of those kids have gone on to attend a recently established watershed-based high school where, among other things, they learn local-appropriate land use techniques. There is no end in sight, and the prospect is now one of ever-deepening experience, rather than one of ever-diminishing possibilities. Our long slow systematic look at the landscape has revealed the tremendous vigor with which nature heals itself, and some of the wonderful logic and time-sense of natural succession is now available to us.

It is part of the process of recovery that we gain a new and deeper perception of home. Before this we lived on parcels, on acreages; now we are invited to live in watersheds and ecosystems, rivers and streams, mountains and valleys. But our emerging identities, fragile and unsure, are bruised and buffeted by forces which reside both within and without our chosen region of effect. The success of state agencies in appearing impermeable; the seeming venality and single-mindedness of industrial resource extractors; the demanding tedium of the court system: together they combine to make a clamoring immediacy which threatens to obscure our purpose — which is transformative — and divert our energies toward the perpetual claims of crisis response.

It is also part of the process of recovery that we learn the things we need to know to live in places. Our crisis response becomes more effective as we come to know more about our particular places than *they* do. But it is likely that our most important and effective contribution to the solution of the puzzle of humans on this planet will be to develop resource-related industries which are *restorative*, which tend to improve air and water quality, biodiversity, and soil fertility, as organic gardening and farming already tend to do. Two examples are emerging in our own Alta California back yard.

Along the northern coast of California, salmon fishing has been a major source of protein and income as long as humans have lived here. By the late 1970s, the combination of growing fleet and dwindling salmon runs had many fishermen casting about for a new occupation. But the number of small one- and two-person trollers kept growing, and among the new members of the fleet were younger fishermen who were to become leaders in an exercise in consciousness-raising for the industry. In a very short time, these notoriously independent operators were to learn a new and reciprocal relationship to the resource which would allow them to continue in business.

In the past, the very independence of fishermen had worked against them. As individuals, they tended to be the best — if not only — source of local knowledge about aspects of the salmon life cycle. But as a political entity, fishermen were most likely to do little more than lobby the government for the least amount of regulation, and fight the annual fight for the best price they could get for their catch. As the catch decreased each year, the fishermen were seized by a mood of deep pessimism. It seemed to many as if the Pacific salmon was doomed to the same sort of depletion that had eliminated the Atlantic salmon as a commercial species. Some argued that no course of action was available but to take all the fish they could until the fishery was gone.

Nat Bingham had spent a winter on Big River nursing a batch of hatchery eggs through California's first stream-

Nat Bingham convinced commercial salmon fishermen to tax themselves in support of stream-repair projects that have helped to restore the salmon and preserve the livelihood of the fishermen.

Brad Matsen/National Fisherman Magazine

It is inevitable, for several reasons, that ecological restoration will take its place on the national agenda of the United States.

William T. Follette

side hatchbox, and he had been impressed by the results. Bingham began to argue that if the fishermen would involve themselves in the fate of the freshwater habitats of the salmon, in the watersheds critical to the reproduction of the fish, it might be possible to imagine a sustained fishery.

One by one, local marketing associations began to line up behind Bingham and the other Young Turks, sometimes by margins of no more than one or two votes, to tax themselves toward a fund that would be invested in salmon enhancement and habitat repair projects — often run by the fishermen themselves. Today, Nat Bingham has been president of the Pacific Coast Federation of Fishermen's Associations for seven years; and that group has spearheaded a Salmon Stamp Program which generates in excess of one million dollars per year for salmon enhancement work in California. In 1988, the salmon fleet had its largest catch in over 40 years. A good part of the credit for that must go to small salmon restoration programs conducted up and down the coast by fishermen's groups and watershed groups alike.

In those same watersheds, natural processes are deeply engaged by yet another set of workers, the loggers. If the commercial fishery was able to develop restorative relationships to its work, could the timber industry do the same? The answer, of course, is yes, and always has been. Though seemingly rare, the presence of small- to medium-size timberland owners managing for sustained yield *and* wildlife and watershed values is as much a constant in the history of timber production as the more noticeable

(and widespread) operators who try to translate whole biomes into cash.

One logger on the neighboring Eel River has taken the notion of sustained yield one necessary step further. Jan Iris, of Wild Iris Forestry in Briceland, California, starts with land cut over within the last forty years and has as his goal restoring forests to their previous diversity and productivity — steady-state, climax, mixed-species Coast Range forests. In the process, it looks like he will be able to make an acceptable income as a timber producer, while rarely harvesting an old-growth redwood or Douglas fir, both now rare enough that every harvest is an object of contention. Since it comes at a time when loggers have lost large numbers of jobs due to automation, and stand to lose more if the larger mills fail to retool for smaller trees, this innovation has important implications for a widespread regional economy.

At the edge of the once-great temperate rain forests, California Coast Range forests exhibit a wider range of species than their neighbors to the north. The incidence of the valuable redwood and Douglas fir is low compared to the incidence of Coast Range hardwoods like tanoak, madrone and chinquapin. These latter woods have rarely found their way to market due to the long-term, hands-on care required to cure them for lumber. Iris has begun to manage timber for small landholders by selecting mature hardwoods and harvesting them in such a way as to release the young conifers which have been kept small by the shade of the great hardwoods. Second and third generations of hardwood trees are also left in situations where

(Left) Madrone trees are one of several overlooked species being harvested by Wild Iris Forestry. (Right) Jan Iris uses a moisture meter to check on the drying of a tanoak board.

Wild Iris Forestry

they will grow straighter and more quickly than if harvest had not occurred. There follows a two-year curing process capped off by a period in a climate-controlled kiln which represents the largest part of Wild Iris's capital outlay (see p. 58). Wild Iris has had to endure a couple of lean years, but has found a demand for its product that exceeds its rate of production. Now local building-supply operator Bob McKee has invested in a six-headed molding machine and has begun to mill tanoak flooring, another local economic turnover.

Iris can see a future with ''an operator up each little watershed,'' with restorative practices as a goal and a reward. ''It doesn't always have to be cut, cut, cut, chop, chop, chop. You may cut trees in the morning and do stream rehab in the afternoon. A lot of things need to be done.'' What is striking about these innovators is that they gained their analysis and inspiration from personal engagement in restoration work — and this is the heart of my

argument for environmental restoration as a transformative process for humans. (Iris gained his insight into the relationship between silviculture and soil erosion doing stream rehabilitation work in the seventies.) More importantly, these are clear models, right in our midst, of ecological restoration transformed into cultural and economic sustainability.

It is inevitable, for several reasons, that ecological restoration will take its place on the national agenda of the United States. The very flesh and blood of evolution, which is wild ecosystems, may already be so severely diminished that the evolution of large plants and animals can no longer proceed. Most experts agree that this potential exists; biologists Michael Soulé and O. H. Frankel claim we have already reached this point. By the time that experts agree as to how much habitat is enough habitat, it is likely that restoration of wild systems will have become not only an appropriate human activity, but an essential one.

This nation has never given much credence to the appropriate, but it is very good at responding to threats. To the degree that ecological restoration is interpreted as a matter of survival, to that degree the U.S. will respond. In that this time-frame coincides with a period when there will be no undisturbed habitats left to isolate and preserve, this effort must spill over into ''resource areas,'' human-occupied, which make up over three-quarters of the planet.

There is no tradition of extended liability for ecological damages. The historical perpetrators are not going to be hunted down and fined. Rural areas will not be able to generate the funds necessary to restore themselves and will always need to appeal to one public agency or another. There will be more Superfunds. But if a national effort at ecological restoration is considered in the context of cultural transformation, and as a pathway to it, it may be possible to limit public costs to a single generation or less, by which time restorative economies can begin to pay for themselves, as consumer appetites adjust to biospheric realities.

The people want it and the talent is available. There is an enormous psychic need on the part of North Americans to engage their continent once more, physically and culturally, evinced on the one hand by a surprising explosion in the sales of backpacks, hang gliders, canoes, scuba gear, and mountain bikes; and more to the point, by a proliferation of in-place grassroots restoration groups. The Hudson River. Papago reservations in the deepest Sonora. Inner-city kids working on weekends to find rare prairie grass specimens in Chicago vacant lots. The Land Institute in Kansas blurring forever the distinction between prairie restoration and appropriate agriculture. Make a list of your own.

It is an indication of the environmental movement feeling its way toward ecological activism that some of its most effective leaders are advocating ecological restoration as a national undertaking. With surprising consistency one hears a misspent military budget proposed as a source of funding, and the Civilian Conservation Corps (CCC) of the 1930s as a model of organization. It is appropriate to demand a piece of the military budget for the job — both to give credence to the enormity of the task, and as recognition of the true nature of security. But the use of the CCC as a model is inappropriate. The CCC was raised as a response to massive unemployment during the depression, and activities were selected, in part, for their non-controversial nature — treeplanting, building park infrastructures, firefighting. Large numbers of workers were moved great distances.

The use of CCC as a model for the current discussion gives rise to the likelihood of seeing the unemployed of Michigan and Louisiana put to work restoring natural systems in California, thus ignoring the potential of restoration work to teach local inhabitants the things they need to know to live integrated lives. If the nation embraces a massive effort in ecological restoration which disregards its potential for social transformation, then, no matter how many trees are planted, species enhanced, or people employed, it will have missed its real goals.

Petrolia High School students release yearling silver salmon into Mill Creek, a tributary of the Mattole. This is done on the new moon in April. Recent studies have shown that salmon smolt on the new moon, a change in their endocrine systems that makes them require salt water to continue living. The dark of the moon also helps these three- to four-inch fish escape predation on their trip downstream to the sea.

Old Growth Forests

Responses to ocean pollution and acid rain require a different level of complexity than is available in my watershed or mountain range. And there are *some* jobs the locals will always be happy to turn over to men dressed in isolation suits who are bossed by Ph.D.s in the capital. Like toxic dumps. Like worn-out nuclear plants. Nevertheless, a very large part of the planet is organized into the visible, and the understandable.

One clear function of an ecological restoration movement is to treat symptoms of habitat decline which are other than human — diminution of species and genetic diversity, deforestation, soil erosion. I am arguing that an equally important goal is to provide individuals and inhabitants the clear experience of themselves as functionally benign parts of living systems. The cumulative effect of these experiences is the transformation of social and economic institutions. As the true nature of environmental decline becomes apparent to more and more people, the desire for social transformation will become more widespread. In the best of all possible worlds, governments will attempt to serve these desires.

Thinking about national approaches to ecological restoration provides an opportunity to broaden the scope of our search for organizational models. One good place to search is in the organization of nature itself. I would like to suggest three principles which we might keep in mind as the mandate to engage in recovery grows:

*Approach the planet as the planet reveals itself.* Ecological restoration must be approached contextually, bioregionally, within the boundaries of natural systems, ecosystems and watersheds. This fairly obvious principle is sometimes obscured by the fact that we do most political things entirely in the context of abstract, artificial boundaries. The absence of alligators cannot be treated in the state of Washington, nor the need for redwoods served in Arizona.

*Human populations of natural areas are necessary participants in local ecological recovery.* Ecological restoration deals with real plants and animals in real time in real places. Real places are not uniform, but break down immediately into a wonderful array of microsystems and microclimates, each of which mediates and modifies toward eccentric behavior on the part of living things within them. These behaviors are the very expression of recovery, and information about them generally resides nowhere else but in the experience of locals. This vernacular science, along with the generalized monuments of legitimate biology and geology, creates the most effective and cost-effective strategies for site-specific habitat repair. Benefits go both ways. You can't treat the fish population in Bear Creek without knowing its current status. Given only one choice, will you mount a three-year study or drop by and visit with the avid angler who has lived for fifteen years directly above the banks of Bear Creek?

*Natural regions exist in time.* One pass through with a government crew isn't going to do the job. The residents will remain in place after the government has come and gone. If the restoration program has been structured so that problems are defined and decisions made by inhabitants with the counsel of technicians, and if much of the work has been performed by local people, especially young people, then a population will remain whose identity has been extended to include their habitat. They will have the skills to maintain equilibrium with the changes inherent in all natural succession. They will have the will to defend the place against further violations. And they will begin to invent the styles of resource development appropriate to the long-range survival of their places, and thus of themselves.

They will have become participants in the planet's recovery. ■

## Nyle Dehumidification Lumber Dryers

Here's a voice from the woods: ''We're a resource colony up here in Northern California. In the old days you needed an army and generals to invade foreign lands before you could extract their resources. But today the United States, with the strongest army in the world, is giving up its resources voluntarily.'' The voice belongs to Jan Iris, whose small-scale Wild Iris Forestry is described in the preceding article. Jan logs native hardwoods that big mills won't bother with, dries the lumber in his kiln, and mills and markets the results. There is a six-month waiting list for his flooring. His large-scale competition is from the Japanese, who buy whole logs from Amer-

ica, ship them to Japan, saturate them with steam and peel entire hardwood logs into 1/64th-inch veneer which is glued onto plywood, shipped back to the U.S., and sold for interior paneling.

The machinery that runs Jan's kiln is manufactured by the Nyle Corporation. Begun as a refrigeration company, it has been making dehumidification dry-kilns since 1978, for drying lumber, fish, leather, peat moss, and fruit. These kilns are basically air conditioners modified to remove water. Nyle is working now to develop a clothes dryer that will look like the ones at the laundromat, but will take nickels instead of quarters.

The reason is dehumidification. A normal clothes dryer, or lumber kiln, introduces heated air which absorbs moisture, which,

with all of the heat, is vented to the outside. With a dehumidification kiln, the heated water vapor is cooled to the dew point, condensed to liquid, and removed, while the heated air is retained and recirculated.

Jan's kiln is in a 20x18-foot barn that can hold 6,000 board feet. A conventional kiln at a large mill holds ten to twenty times that amount, but, in part because his dehumidification kiln is more efficient, Wild Iris Forestry is a viable small business.          —Richard Nilsen

**Nyle Dehumidification Lumber Dryers: $1,600-$500,000.** Information **free** from Nyle Corporation, P. O. Box 1107, Bangor, ME 04401; 800/777-6953.

---

## The Redesigned Forest • Forest Primeval • From the Forest to the Sea

The more we begin to understand how forests function as ecological systems, the clearer it becomes that modern forestry is akin to mining, not resource management, and precludes any hope of sustainability. In the forefront of scientists doing the fieldwork to back up these kinds of assertions is Chris Maser. He writes books in a style equivalent to an ecological food web -- a careful look at process via a tangential combination of hard science, history and humanistic psychology.

*The Redesigned Forest* contrasts the ecological needs of a forest with current short-rotation forestry practices. If you want to understand his basic argument, start here. *Forest Primeval* is a biography of an ancient forest in Oregon, from its beginnings in the year 988 up to the present. *From the Forest to the Sea* is an example of the fieldwork Maser was doing at the Bureau of Land Management before he left to become a private consultant. It examines what happens to the biomass of fallen trees, in the forest, in the watershed, and even on the seabed miles off the Oregon coastline.
          —Richard Nilsen

•

When we think of Nature's forest as a commodity, we treat it like one. Because

we treat it like a commodity, we are trying to redesign it to become one. We take a system designed by Nature to run in 400- to 1,200-year cycles and attempt to replace it with recurring cycles of only the first 80 to 120 years. We do not see the forest. We are so obsessed with our small goals

that we . . . redesign the forest with an instability that cannot be repaired with fertilizers, herbicides, or pesticides.

•

Decades of scientific research have concentrated on every possible cause of forest decline *except* that it might be the direct result of intensive plantation management based on ignorance of forest processes. The forests of central Europe are now dying . . . and we hear much about acid rain. . . .

I agree that atmospheric pollutants may well be playing a role in *Waldsterben* (the dying forest). Yet I cannot help but wonder how the cumulative effects of a century or more of intensive plantation management and use may have strained the forests of Central Europe, and thus predisposed them to the ''*Waldsterben* syndrome'' we see today.          —*The Redesigned Forest*

•

The Douglas-firs in the burn near the spring are now 250 years old and between 200 and 220 feet tall. . . .

By now, 1237, their crowns have lost much of the pyramidal form and have become rounded or somewhat flattened. And by the time they are 450 years old in 1437, their crowns will be cylindrical and resemble a bottle brush (albeit one missing many bristles). Their trunks will be clear of branches from 65 to 130 feet above the ground. . . .

The upper surfaces of the large branches will become covered with perched, organic ''soil'' several inches thick that will support entire communities of epiphytic plants (plants growing on other plants, in this case primarily mosses and lichens) and animals. Large branches will become the home for myriad invertebrates, as well as some birds and a few mammals. . . .

A single ancient tree can have over 60 million individual needles that have a cumulative weight of 440 pounds and a surface area of 30,000 square feet, or about 1 acre.          —*Forest Primeval*

**The Redesigned Forest**
Chris Maser, 1988; 234 pp.

**$12.95** ($14.45 postpaid) from R. & E. Miles, P. O. Box 1916, San Pedro, CA 90733; 213/833-8856 (or Whole Earth Access)

**Forest Primeval**
Chris Maser, 1989; 282 pp.

**$25** ($29.50 postpaid) from Sierra Club Store Orders, 730 Polk Street, San Francisco, CA 94109; 415/923-5500 (or Whole Earth Access)

**From the Forest to the Sea**
Chris Maser, Robert F. Tarrant, James M. Trappe and Jerry F. Franklin (Stock #001-001-00642-4) 1988; 153 pp.

**$15** postpaid from Superintendent of Documents, Government Printing Office, Washington, DC 20402-8325; 202/783-3238

Bald eagles choose to perch on mudflat objects, such as driftwood, rather than on the tidal substrate. *—From the Forest to the Sea*

Watershed
Rehabilitation
at Redwood
National Park

*BY KAREN GRIFFIN*

SINCE the time when dinosaurs roamed the planet, redwood forests have thrived in environments such as those now found along California's north coast. Recently, however, logging and other land uses have drastically transformed these lands. Old-growth redwood forest ecosystems have been altered to the point that there are serious concerns they can again support a redwood forest community as complex and naturally diverse as those that were cut for lumber. Left to its own devices, deforested and eroding land should slowly recover, but not to its former state or stable condition, at least not during our children's or their children's lifetimes. At Redwood National Park an effort of unprecedented scale is taking place, designed to give nature a head start in restoring logged land. Lessons learned during this restoration program could be useful to those interested in helping to heal other wounded, but still living, lands.

Redwood National Park, established in 1968, originally incorporated 58,000 acres of old-growth coastal redwood forest and associated streams and seashores. Primeval redwood groves are found along the coastal strip and streamside alluvial flats in the lower one-third of the Redwood

Creek basin. In hindsight, what the 1968 park design lacked was a watershed perspective. Even though the groves of the world's tallest trees were protected from being logged, they weren't protected from the effects of logging upstream.

Adverse timber harvest impacts are magnified by the setting. The park's proximity to the boundaries of three crustal plates has resulted in myriad faults and active uplifting. Bedrock is composed of soft, fine-grained, marine sedimentary rocks, often sheared beyond recognition by tectonic processes. The area receives 80 to 100 inches of rain annually, mostly during intense winter storms. Erosion rates are naturally high. Over geologic time, a dynamic equilibrium was maintained between uplift and erosion rates. Streams evolved to store and transport sediment loads carried to them. The redwood-forest biological community thrived in this environment. ▶

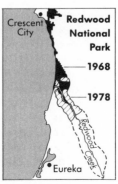

*Karen Griffin is a physical-science technician in the watershed rehabilitation program at Redwood National Park in Northern California. The worst thing I've heard anyone say about the mammoth restoration efforts going on there was from one fellow who groused, "What the Feds are doing is proving that if they spend enough money then, by God, they can do restoration." I suspect he may have had a case of budgetary jealousy. The backhoe in the photo above is from a 10-year-old project. Nowadays, they have been replaced by giant excavators with buckets big enough to scoop up a Volkswagen bug. I think the Park Service is also revising for all of us the old adage about what an ounce of prevention is worth — in hilly terrain with heavy rainfall, it is worth tons of cure.*
*—Richard Nilsen*

Restoration before, during, and after:
A logging road and culvert (left) is removed (below). The same site (right) three years after road removal. The crews at Redwood National Park make an interesting distinction — they refer to all their work as rehabilitation, and leave the ''restoration'' to Mother Nature.

But just outside park boundaries, this long-standing, natural equilibrium was decisively tipped in favor of erosion. Earthmovers cut logging roads to get at valuable trees. Soils and hillslope drainages were disrupted as bulldozers constructed earthen mounds upon which loggers dropped brittle redwood trees. Logs were mostly dragged down-slope by tractors to loading areas, but if slopes were too steep for bulldozers, logs were dragged by cable to hilltop areas where log trucks were loaded for trips to sawmills. Steep, mountainous areas devoid of vegetative cover lay bare and unprotected against intense rainfall. In critical places 10 to 20 inches of rich topsoil, the active biotic layer, were flushed away down gullies. During major storms, water, mud, and logging debris spewed down drainages too small to handle the newly diverted runoff. Fill that had been pushed into streams to build roads washed out. The frequency of landslides and earthflows accelerated dramatically throughout the basin.

Downstream, in the park, giant redwoods toppled as sediment accumulated and floodwaters undercut stream banks. Fish spawning gravels were buried under sediment and summer water temperatures climbed, while the cutting of marketable timber continued to threaten park resources. Following a hotly contested conservation battle, the Congress enacted legislation, signed into law by President Jimmy Carter in 1978, to expand the park by 48,000 acres. National Park Service management policies typically dictate that parklands be preserved in their present state for future generations. At Redwood, however, Congress authorized the National Park Service to rehabilitate the more than 36,000 acres of clear-cuts added to the park.

Behind park expansion was the conservation community's enlightened philosophy of looking at an entire watershed's dynamics, from ridgetop to ridgetop and from headwaters to the ocean. Reaching the goal of preserving the entire watershed fell short of the desired ideal. Political and economic constraints at the time made it impossible to include the upper two-thirds of the basin.

Today, logging continues in the upper basin while the park monitors impacts on downstream park resources.

A watershed rehabilitation program was undertaken on the lands added to the park. Unlike the sites of most restoration efforts taking place today, this large tract of land is held by a single landowner (the public), without competing goals or multiple consumptive uses. The park's watershed rehabilitation team is made up of wildlife biologists, geologists, hydrologists, botanists, and archaeologists. When the program first started in 1977, state-of-the-art watershed rehabilitation consisted primarily of small-scale, labor-intensive techniques. Many of the check-dams, wooden terraces, flumes, and wattles installed in 1978 and 1979 are functioning today. Some washed out during the first big storm because of design failures or because they were located in remote areas and could not be routinely serviced. Although these structures are effective in slowing erosion, they are not designed to remove the massive potential sediment sources still perched on hillsides.

The sheer volume of soil and debris left from the logging called for earth-moving strategies that far surpassed anything attainable by a manual labor force. Some environmentalists cringed, but bulldozers and excavators which had created the mess in the first place were called back to clean it up. Studies comparing heavy-equipment costs versus labor-intensive treatments fueled a major change in the watershed rehabilitation program. There was a shift towards using heavy equipment to more completely excavate and reshape soil removed from drainages and road fills.

Eleven years ago, when the restoration began, there were 300 miles of haul roads and 3,500 miles of skid trails criss-crossing drainages. About half of those miles have been rehabilitated. Areas have been surveyed to determine which roads have failed and to prioritize those posing the greatest erosion threats. Each year, new areas are tackled. The first step is to interpret historical aerial photos for past land-use practices, such as the age of roads and the size of the logged tracts. Drainages and gross geomorphic features are studied in the office before fieldwork begins. Original stream courses must be identified before a road is ''put to bed'' and natural drainage patterns re-established. After this scientific, left-brained, data-gathering phase comes a more right-brained, artistic prescription phase. A complete excavation must be intuitively visualized before the first tractor rolls. The lay of the land guides this process. Once the rehabilitation site

bulldozers equipped with hydraulically operated, giant ripper teeth. Drains are placed every 100 feet to take water out of the former roadbed. After the heavy equipment moves out, hand crews complete the work. Bare earth surfaces are blanketed with straw mulch, or in some cases with compost. Native conifer seedlings are then planted.

Sites found falling apart in the winter rains, and reshaped by heavy earthmovers the following summer, are covered with a lush green mantle of nitrogen-fixing alder and wildflowers. The results of this work give hope that nature can quickly heal wounded landscapes if unimpeded by excessive erosion and repeated exploitation. Watershed rehabilitation is expensive. And it takes many human lifetimes for an old-growth forest to regrow and mature. Successful restoration of decimated lands should not be used as a justification for further destruction on this scale. Restoration should teach us to appreciate the land's complexity and the importance of good stewardship as a more effective management strategy for the future. ∎

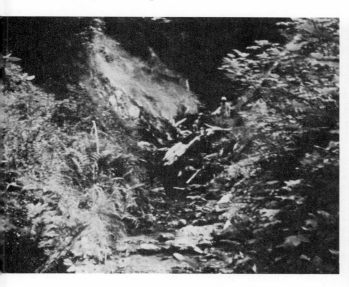

prescription is complete, a survey crew measures dimensions to calculate how many cubic yards must be moved. Work areas are laid out using stakes and flags.

During the dry summer season, bulldozers, hydraulic excavators, and dump trucks move earth until the physical landscape is once again in a stable configuration. Fill from the outboard edge of some road sections is stacked against the inboard cut-bank to completely recontour hillslopes. Topsoil is recovered from below the road fill and replaced on reconstructed hillslopes to help foster revegetation. Some road sections have seeps or springs that could saturate replaced fill and cause failures. Here, earth must be exported to a ridge or other stable fill site.

The process of road construction is reversed during stream-channel excavation. Culverts are removed and thousands of cubic yards of fill that were pushed into the channels to make flat road surfaces are removed, until the original stumps and stream-bottom deposits are encountered. Less erosion-prone roads are decompacted by

A redwood forest ecosystem. The vegetation is lush right to the high-water line, and the forest canopy provides shade to keep the stream cool enough in summer so that small salmon can thrive.

Each summer, a field seminar on watershed rehabilitation is presented at Redwood National Park in cooperation with Humboldt State University. Also, a number of park-published technical reports evaluate watershed rehabilitation progress and related research and monitoring programs. Many are available upon request. For more information, contact: Superintendent, Redwood National Park, 1111 Second Street, Crescent City, California, or telephone 707/464-6101.

# TURNING BACK THE CLOCK IN THE FLORIDA EVERGLADES

*BY RANDY WILLIAMS*

**T**HERE IS A SAGA UNFOLDING in the state of Florida that will lift the hopes of the environmentally conscious worldwide. At the eleventh hour — maybe even the eleventh-and-a-half hour — under the visionary leadership of the state's governor, Bob Graham, projects were initiated that are staving off the depletion of fresh-water resources and the destruction of hundreds of thousands of acres of irreplaceable wildlife habitats in and around the Everglades. Water is the lifeblood of this area. It has always been that way.

Only within the last 10 to 15 years has there been widespread understanding of how the Everglades functions and of its impact on the water resources for all of south Florida. There are really only two distinct seasons in this sub-tropical environment, wet and dry. Almost all of the fresh water for this region begins as rain; 60 inches of rain annually saturates the spongelike peat soil and overflows the rivers and lakes. Then it starts its slow shallow trickle south, only inches deep, dropping just one foot in twenty miles on a path 40 to 50 miles wide, and finally slipping into Florida Bay. This sheetflow, as it is aptly called, allows fresh water to filter through the sawgrass and peat, sand and limestone to recharge the two great subterranean aquifers, the Florida aquifer in the north and the Biscayne aquifer in the central and south regions, which are the only sources of drinkable water in Florida. Countless species of plants and animals depend on the timely arrival of this water to enrich food and habitat resources. During the dry season, the process reverses itself and the soil slowly releases its water reserves, maintaining this delicate balance between flood and drought.

The sheetflow is dependent upon the rains; but there is

strong evidence that the rain is dependent upon the evaporation of the massive sheet flow, a closed cycle that is so simple yet so very delicate. In the years since the natural sheetflow has been dramatically altered by levees and canals, the rainy season has also been less dependable. Water levels in the great aquifers have dropped at an alarming rate, and dryer conditions contribute dramatically to the severity of fires that burn not only vegetation but also the precious peat soil itself. Additionally, colder weather and killing frosts have crept farther south each year. As more and more fresh water has been used from the subterranean aquifers, there has been a steady encroachment inland of salt water. Once salt water contaminates a well, it can never be converted back to fresh water. Historically, the fresh water was so plentiful in coastal regions that passing ships could stop offshore, in the Atlantic, and replenish their water kegs from fresh water springs bubbling up through the salt water.

From Florida's earliest days as a state, the Everglades have not been fully appreciated. The first legislature in 1845 characterized them as ''wholly valueless'' and called on Congress for assistance in draining the swamp. Ironically, Governor Graham's father Ernest Graham, a former

---

*Restoration work can involve a lot of un-doing, in this case the monomaniacal flood control handiwork of the U.S. Army Corps of Engineers. Wetlands are very fragile; perhaps this is why south Florida hit the wall, environmentally speaking, sooner than some other places. The state risked jeopardizing not only its largest industry, tourism, but its supply of fresh drinking water as well. Florida is responding with policy changes and action at the highest levels, exactly what is required for a problem this extensive.*

*Randy Williams is a photographer (a refugee from corporate audio-visual work) who resides in West Glover, Vermont, where he also writes, does carpentry and teaches art. For more information about the Everglades Coalition, contact Theresa Woody, 1201 N. Federal Hwy., Rm. 250-H, N. Palm Beach, FL 33408.*
*—Richard Nilsen*

captain in the Army Corps of Engineers, was one of the proponents of a major drainage policy in the 1930s. Severe floods in 1926 drowned 1,836 people and another in 1947 caused $2.3 million in damages. The devastation gave the drainage advocates the ammunition they needed to persuade Congress to authorize the Army Corps of Engineers to develop a course of action for flood control. The result was a massive, ill-conceived ditch and dike program of proportions unequaled in the US. Throughout the 1950s and '60s more than 1,500 miles of canals and ditches were gouged out systematically throughout the Everglades' wetlands. Around Lake Okeechobee, the second largest fresh-water lake within the United States, 106 miles of 35-foot-high dikes with flood-control gates were constructed, effectively stopping the natural sheetflow. The Kissimmee River, with its 97 miles of meandering oxbows, had for centuries supplied naturally filtered water to Lake Okeechobee. But the Corps transformed the river into a 54-mile straight-shot channel. Additionally, dikes were constructed along the Tamiami Trail, US Route 41 cutting through the Everglades from Naples to Miami. All of this was done under the guise of a ''flood control program.'' And all of this plumbing did its job. The swamps and wetlands started to dry up, and for the ever-hungry land developers there were visions of sugarplums. Indeed, in the 1960s the notorious subdivision, Golden Gate Estates, was started on the northwest edge of the Everglades, digging 180 miles of canals to drain 173 square miles of wetlands. But with no promise of water or sewer

service and a fierce resident mosquito population, only a small area has been actually developed. The drainage, however, lowered the water table more than two feet.

Throughout the last three decades the influx of people into Florida has been phenomenal. It is estimated that an average of 800 people a day are moving to the state, and with that growth comes an ever-increasing demand for space and water. Sewage and solid-waste disposal are a major concern. Agribusiness, hundreds of thousands of acres of sugarcane, vegetables and citrus to the south of Lake Okeechobee, also take a great toll on water resources, using nearly half of the total water consumed and dumping millions of pounds of toxic pesticides into the soil yearly. Much of the excess high-nutrient irrigation water is back-pumped into the lake. Dairy cattle along the Kissimmee River canal are contaminating the water with phosphorus-laden manure which shoots right down into Lake Okeechobee with little opportunity to decompose and to be absorbed into the soil. The nutrient-rich water has spawned algae and other vegetation that is steadily robbing the lake of oxygen.

The real problem was that everybody wanted their share of the water where and when they needed it. The state established distinct water-management districts to deal with the dispersal. Schedules were set up and water was delivered to fill almost every need. If, during the rainy season, flooding seemed imminent, excess water was unceremoniously dumped through canals into the ocean.

►

The Kissimmee River used to be 97 miles of meandering oxbows. In the 1960s the Corps of Engineers transformed it into the 54-mile Kissimmee Canal C-38. The wetlands dried up and dairy cows replaced wading birds, whose populations have decreased 90 percent in the last thirty years.

Canal retro-fit: One of three steel weirs installed in the upper Kissimmee Canal (top). The weirs constrict the flow enough to back water up into the old oxbows (above). This filters the water, recharges the aquifer and restores wetland habitat.

Unfortunately, judicious distribution proved to be a hopeless task, one that turned into a political power struggle, and the environment ended up at the bottom of that barrel. Natural water levels in and around the Everglades fluctuated drastically according to domestic and agricultural needs outside of those areas. Consequently, breeding and nesting areas were either too wet or too dry, and dependent animal populations steadily declined.

All of this "management" of the water and wetlands has devastated much of the natural ecosystem in and around the Everglades. It has been estimated that over the last thirty years, the wading-bird population has dropped by 90 percent. This is largely due to unnatural fluctuations in wetland water levels that flood nesting sites. Residential and agricultural expansion has pushed wildlife into smaller and smaller areas. This loss of habitat is the main contributing factor threatening already endangered species such as the Florida panther.

**B**OB Graham grew up on the edge of the Everglades and witnessed firsthand the evolution of the problems it now faces. His family's dairy farm was tenuous business during the 1940s and '50s. Wetlands and periodic flooding were obstacles to be overcome, not compromised with. Growth and prosperity were the good guys and the swamp the bad guy, waiting ominously just on the other side of the ever-expanding fence line. But Bob proved to be a different breed of Graham, realizing that nature was more than just a menace to progress. "I guess I am less likely to agree with my dad that the way to solve a problem is to alter nature." Bob reflects, "My experience is quite typical of others who have spent a lifetime in South Florida: incremental, unnoticed change, then a jolting awareness of what had happened." As a state senator he gained a reputation for environmental conservation. Elected governor in 1978, he started rebuilding the state's political structure to loosen the deathgrip that agricultural and development interests had on water management. In 1986, Bob Graham was overwhelmingly elected to the U.S. Senate, where he continues to champion the cause of water and environmental rehabilitation in Florida. His successor to the governor's post, Bob Martinez, has not only continued the environmental program but expanded it to include the cleanup and protection of Lake Okeechobee and to regulate the development of citrus land conversion in southwest Florida.

Even though the state enacted laws protecting various aspects of the water and land usage, they were not broad-based enough to reverse the damage that had been done. Environmentalists and conservationists had organized their efforts and, with the public sentiment shifting in favor of wildlife conservation, politicians were made aware of the dire situation that had been created. In 1983 Governor Graham initiated a bold new environmental program that would preserve and expand what remained of the state's fresh water resources and revitalize the diversity of wildlife habitats in and around the Everglades. The Save Our Everglades program took a comprehensive, holistic approach to a very complex problem. It was, and is, at-tempting to undo many of the short-sighted environmental travesties that have taken their toll on this delicate ecosystem in the name of progress and growth since the early 1900s. In the words of Governor Graham, "Our goal, by the year 2000, is to have the Everglades look and function more like it did in 1900."

Initially, there were six major elements to the program: *1.* To reestablish the values of the Kissimmee River and its floodplain; *2.* Restore and protect the 95-square-mile area north of and adjacent to Water Conservation Area 3 in Palm Beach County; *3.* Manage the deer herd in Water Conservation Area 3 just west of Fort Lauderdale; *4.* Improve the hydrology along the Tamiami Trail and Alligator Alley, which connects Naples and Fort Lauderdale; *5.* Restore the natural hydrology of the Everglades National Park; *6.* Protect the Florida panther and other endangered wildlife.

Graham realized that in order for such a grandiose plan to have any chance of success, he would have to appeal to the hearts and minds of the people of Florida. And that he did, with the support of local, state and national environmental organizations. He took his message of conservation and preservation to the news media. Throughout the program the people have been well informed of the progress and how much its success will affect their future. They have also been apprised of the practical aspects of saving the Everglades and other nature habitats. In addition to providing drinking water, the Everglades system is a major tourist attraction. Tourism is Florida's most important and profitable industry, bringing in ten times the dollars that agriculture does.

Governor Graham secured the participation of over 20 state and national environmental groups, including National Audubon, Sierra Club and the Wilderness Society, in an effort to overcome the many public, legal and political obstacles that such a controversial multifaceted program encompasses. In 1984 these organizations formed the Everglades Coalition to coordinate efforts on the state and federal levels in promoting action from a maze of government agencies. The Coalition lobbies state and federal officials, including the president, testifies for programs before congressional hearings and subcommittees, initiates litigation if necessary, and promotes public awareness of environmental issues through each organization's publications and newsletters. Thanks to the tireless efforts of environmentalists such as Marjory Stoneman Douglas, whose 1947 book, *The Everglades, River of Grass*, eloquently predicted the current crisis, public awareness and political action are beginning to turn back the clock for the Everglades.

Much has been accomplished in the six years since the program's inception. Although the vast majority of energy is spent on political and legislative administrative fronts, the tangible results of the program are quite impressive as well. Three steel weirs, or partial barriers, have been installed in the northern section of the Kissimmee canal. The weirs constrict the flow of the canal, allowing water to flow back into the original river channel. This project has restored about 12 miles of the river and 1,300 acres

Florida Game and Fish Commission

From Naples to Fort Lauderdale, State Road 87 is known as Alligator Alley. Its replacement by Interstate 75 includes thirty-six underpasses like this one, strategically placed to permit water flow and provide safer crossings for wildlife.

Betsy Day

Somewhere between thirty and fifty extremely reclusive Florida panthers are left in the entire state. This one was hit by a security vehicle at the Fort Myers airport. Nursed back to health by the state's Panther Recovery Team (and fitted with a radio collar), it is being released at the Corkscrew Swamp Sanctuary, owned and operated by the Audubon Society.

of river marsh and valuable wetland habitats. During the rainy season, thousands of acres of the historic floodplain are again flooded. Of the 50,000 acres needed along the Kissimmee to complete the project, 21,500 acres have been procured by the state.

The increased concentrations of phosphorus and nitrogen in Lake Okeechobee over the past twenty years have spawned the growth of algae and other vegetation that deplete the water of oxygen. This eutrophication threatens the entire food chain in the lake. New waste-management practices by dairy farmers along the Kissimmee and the reduction of the backpumping of nutrient-rich water into the lake from agricultural interests south of the lake have greatly decreased the influx of these contaminants. Although the lake is under a strict monitoring program, results from these projects are going to be slow. Other cleanup measures are in the legislative stage. Bass fishermen have organized to assist in the planning and promotion of water quality on the lake.

Land acquisition is a big part of the master plan. In the water conservation areas south of Lake Okeechobee and west of a strip stretching from West Palm Beach down to Miami, the state has been negotiating to buy nearly 90,000 acres of private land. This area, which has suffered from overdrainage and peat fires, would be converted back to its wetland state to help replenish a more natural sheetflow supply to the Everglades National Park. A 3,700-acre tract has been approved for use as a biological filtration system for runoff water entering the area. Undeveloped land in the Golden Gate Estates is also being purchased, to be converted back to a wetland environment. An innovative proposal by the U.S. Department of the Interior, in conjunction with Collier Enterprises and the Barron Collier Company, would exchange about 108,000 acres of wetlands for 68.4 acres in downtown Phoenix, Arizona. A similar swap, land that had been slated for the ill-conceived Aerojet Port in the southeast Everglades, was made for some federal property in Nevada.

The conversion of Alligator Alley to Interstate 75 with the inclusion of strategically placed water and animal underpasses is well underway. The underpasses will allow a more natural water flow into the Everglades National Park.

The highway runs east from Naples through the Fakahatchee Strand and the Big Cypress National Preserve, which

are the primary ranges for the endangered Florida panther. In 1982 a vote by Florida schoolchildren selected the Florida panther as the state mammal. Because the panther's natural range crisscrosses over major traffic arteries, several animals are killed or injured each year. Road signs and speed zones have been installed to alert drivers.

In April of 1989 Senator Graham introduced the "Everglades National Park Protection and Expansion Act" which, if passed, will add nearly 110,000 acres to the eastern edge of the park. *[The bill passed and was signed into law in December, 1989 —Ed.]* Experimental water-release programs are attempting to restore the hydrological balance to the park, which is also part of the range of the panther. With less than 50 of these reclusive animals remaining, a Panther Recovery Team has been formed by the Game and Fresh Water Fish Commission to gather data and study their habits. Twelve cats, either captured or found injured and nursed back to health, have been fitted with radiotelemetry collars and released. Now their movement can be monitored from the air. The Commission is also conducting a captive breeding program.

The Everglades is a precious place of unique beauty and wonder. It is a self-contained ecosystem that, given the chance, can serve man and nature, but it has a delicate vulnerability. It presents, in microcosm, all that is good and bad about the way we are treating our planet. We can alter and exploit the environment only so much before the systems collapse. The Save Our Everglades initiative is a program of long-term preservation. Without it there would soon be no Everglades, no clean fresh water, no agribusiness, and only a few people. It is a program of hope. ■

A neighborhood in ruin —
Beck Street at Fox, South
Bronx, 1978.

Carlos Ortiz

# Restoring Cities From The Bottom Up

## A Bi-Coastal View From The Street

### BY MIKE HELM AND GEORGE TUKEL

*Say the words "urban restoration" and most people think of fixing up old houses. Urban gentrification has spawned a whole industry during the last twenty years, one well served by magazines like* The Old-House Journal. *This movement began as a reaction to the urban renewal of the 1960s, when entire neighborhoods of old houses were simply bulldozed. Today it is not the houses that are in jeopardy. As a friend of mine says, explaining why he quit the historic preservation business in San Francisco, "I realized a while ago that there were enough Gucci-shoed lawyers to protect every last Victorian in San Francisco."*

*It is the urban public spaces that are now in need of restoration — the sidewalks, schools, parks, and playgrounds. Because these environments are public, they are much more vulnerable to abuse and destruction than houses, which have owners to protect them.*

*Mike Helm is a writer, and also publishes City Miner Books from his home in a warehouse district in Berkeley, California. George Tukel has helped neighborhood groups design pocket parks in New York City. He now lives in Oakland, California, and is a partner with Mike in City Miner Salvage.* —Richard Nilsen

WHEN people talk about how best to restore cities, the debate is usually framed within the boundaries of two ostensibly different perspectives. On the one side are professional environmentalists, on the other advocates of growth. Reading the daily paper, and watching the evening news filled with sixty-second stories, these two camps appear as antagonists. Yet a strong case can be made that the Sierra and Rotary Clubs — in terms of their institutional effect — aren't that far apart. Through the planning, legal, and media processes, we believe, they more often than not legitimize each other in ways that ultimately perpetuate the current displaced industrial culture that is destroying our cities. To see how this could be so, let's look a little closer at each of these perspectives.

Typically, environmentally oriented advocates expound upon the need for clean air and water, energy conservation and the development of sensitive waste-management systems. For them a strong imperative is the development of more urban sanctuaries — like green belts and open spaces — devoid of concrete and automobiles, that would bring humans into closer touch with nature's other plant and animal communities. Underlying this vision is the humanistic view that the only way to make cities livable is to minimize the direct and secondary effects of industrialization.

For growth-oriented advocates, urban restoration is centered on successful and expanding businesses and ever more development. They are less concerned with pollution and communing with nature than with making money and creating jobs, to expand the tax base, to upgrade the social services with which to fight major urban problems like unemployment, crime and homelessness. The trickling down of corporate culture is seen as the means to remedy social ills.

In the real life of cities, of course, the record of urban

decay — with its pollution, homelessness, drug problems and general disintegration of the physical infrastructure — proves that neither the environmental nor the corporate view has been sufficient to restore our cities. Even though specific elements from each of these views have validity, they have ultimately reenforced each other as one ongoing pathological program of urban growth and decay. While environmentalists may see themselves as restorationists when they mitigate development plans, we believe that in fact, they have become blind, if not antagonistic, to the alternatives that would restore our cities. More often than not, they wind up — through the mitigation process — legitimizing developers and the top-down structure that has brought us to the current conditions. If urban restoration is to occur, we believe that it will happen only within a framework that is both broader and different from either the corporate or environmental agendas.

The evolution of cities currently has more to do with the short-term profits of developers, constrained by zoning and environmental regulations, than it does with insuring the viability of the diverse communities that make up cities. Our restoration view assumes a fundamental departure from this way of getting things done — not only in terms of a wise and conserving use of resources, but also in planning and decision-making as well. Our view of urban restoration involves a grassroots, street-level emphasis on making cities convivial places to live in by wedding natural and human ecology. Without allowing people to interact more freely with each other and to identify more closely with the natural systems within which they live, it isn't possible to make urban life convivial. Ultimately, there is a connection between an enduring urban culture and the health of the rivers, hills, creeks and bays that surround cities.

Our view of restoration is also traditional. We recognize that cities are about people. An important part of what we mean by conviviality involves nourishing ongoing community vitality. We want to encourage the enjoyment of a festive society. We extend the meaning of conviviality to the art and practice of living together as a community — the breaking of bread among neighbors — within the context of neighborhood character and scale. Toward this end, the memory of a community is important to its identity and can only be maintained by the preservation of its past.

Urban restoration also depends on the re-establishment of the public domain in the form of the ''Commons'' — places for everyday socializing, like mixed-use community parks, or sidewalks and streets designed as meeting places for weekend produce markets, neighborhood swap meets and block parties. Reconnecting urban awareness to natural systems is also necessary. And convivial urban

*Environmentalists have become blind, if not antagonistic, to the alternatives that would restore our cities.*

restoration requires a loosening of the licensing and zoning structures so that all members of a neighborhood can participate creatively. Taken together, these should be the standards for evaluating the incremental changes that occur in the feeling, memory and infrastructure of the community's physical space.

Our view combines what we know of the attributes of flourishing cities and natural systems. It includes the bioregion as an element of cultural identity and economic security, but also scales down to neighborhoods for local autonomy. Neighborhoods are the basic unit of survival within cities and flexibility in committing local resources is the key to adaptability. Said differently, the restoration of cities is a cultural and political process informed by ecology.

*George Tukel:* I grew up on suburban Long Island. My first real exposure to city living came when I was working my way through college in the late sixties. I drove a truck for a friend of mine, and the warehouse I shipped out of was in the South Bronx. If you drive through South Bronx today, you'll find more open and green spaces than in just about any other place in New York City. At the time that I was working there, a good part of the Bronx was being burnt to the ground by a community angry about the lack of control it had over its financial well-being, the general squalor of the neighborhood and the community's destiny. After work I used to watch things burn. Though often hectic, life during that time was never lived inside; it was always hot and on the street. This included eating, drinking, listening to music, socializing, flirting and passing along and embellishing the latest news. City life is street life and the streets have to be alive for the city to be alive. When you walk through gentrified neighborhoods in New York now, there's none of that juice. Prettiness has replaced the magic of people.

Increasingly, newer buildings have replaced the previous meeting places where things used to happen. Everyday areas — the front stoop, small green spaces, the nestled public square, the outdoor market — are all missing. When meeting places are designed into new construction, they seem incidental and are there only as part of some mitigation package that city planners and developers have worked out to allow the developer to have more units. The result is sterile and unsocial and it is not surprising that people don't partake, but find themselves living increasingly isolated lives.

Architects and developers need to rein in their egos, put their so-called ''signature'' buildings on the back burner and start designing with neighborhood people, using successful local precedents to support community.

*Mike Helm:* I grew up in the rural San Fernando Valley and then moved to suburban North Hollywood as a kid

Street fairs, like this one on Telegraph Avenue in Berkeley, California, strengthen community vitality by turning the street into a form of "commons" — a meeting place for socializing. Mike Helm (right) sells the books he publishes.

in the fifties. I didn't get a real feel for urban street life until the late sixties when I lived in south Berkeley. In the early seventies most of my time was spent writing and selling poetry on Telegraph Avenue. A lot of people were hustling a living directly from the street. A couple of hundred people, including myself, still are. What I liked about the scene then was that the sidewalks were alive and people didn't avoid each other with averted eyes. In fact, people watching was one of the major enjoyable activities. We actually took pleasure in checking each other out and, because there were so many of us to share the load, even gave away a little of our spare change. About the only thing that was missing were public toilet facilities like they have in some European cities. It continues to amaze me that we have a whole country dedicated to the notion that bodily functions somehow stop when people leave home or aren't buying something in a store. It's such a basic human need. There are lots of older people and parents with children who understandably don't go anywhere unless they know where the next bathroom is.

One of the things that made this sidewalk society safe and interesting was the large number of street artists, food vendors and pamphleteers hustling on the streets. There were a lot of crazy people too, but in the larger stream of humanity, they were, unlike now, somehow attended to and absorbed. With all the little kiosks dotting the sidewalks, people didn't simply rush past. The sidewalks weren't just the pedestrian equivalent of freeways, exclusively designed to get foot traffic from Point A to Point B. They were something to be savored and experienced. The sidewalk ambience gave people an excuse to hang out, listen to street musicians and even shop. For the vendors

that had something interesting to sell, a little commerce even ensued. Nobody was getting rich, but a lot of otherwise unemployed people were getting by.

*GT:* Right, the streets worked as a commons. By commons, I mean a physical area that no one person or group controls for their immediate self-interest, but a place that everyone benefits from. Streets are commons for cities in much the same way that pastures for grazing were commons in pastoral England and can only survive in much the same way. In the pastoral example, people knew the limits of grazing necessary to preserve the carrying capacity of the land and consequently the community. So all exercised restraint in the number of livestock they grazed. When one member broke the unspoken rule of restraint (by increasing the size of their herd), they not only destroyed the resource base, but damaged the mutual trust upon which community living depends. This dynamic is not unlike the one that occurs when a developer buys up half a city block and puts a huge building on a neighborhood street. Their own private agenda destroys the commons that was previously the streetscape.

I'm interested in the way that streets and buildings can meet to create commons and allow individual participation. What kind of design forethought — even down to the level of locating curb cuts and the dimensions of sidewalks — encourages people to sit outside, or put flower boxes on their windows?

*MH:* The loss of the street for pedestrians has been reenforced by modern planning processes. While I agree with you about the limits of privatization, I also think that

Richard Nilsen

public spaces suffer from monocultural policies. Parks are an example of this. If it were up to me, I'd encourage much greater mixed use of park space. I'd like to see a lot more small-time vendors and street artists running their gigs on the sidewalks that surround the grassy areas. People enjoy and use parks more when something interesting is happening around them.

I'd also like to see wider use of school yards and transportation parking facilities, too. As an example, a rapid transit parking lot, unused on weekends, was "liberated" by direct action in south Berkeley for use as a weekend flea market. After some tense confrontation with bureaucrats, who only saw themselves as transportation experts, the flea market was legitimized and has subsequently evolved into a thriving market commons.

One of the reasons people support flea markets is because such primordial human traits as foraging and bargaining are given sway there. People really don't enjoy a pre-packaged environment. They don't want to be just consumers, they want to participate in and help shape the economic process. You see this especially with recent immigrants from traditional, "underdeveloped" cultures. They demand that you deal with them personally. They want to fondle and caress what is around them — see if the fruit is really ripe.

The way to deal with social ills is the way the rest of nature does it, through dilution. Rather than congregating all of our homeless in one or two parks, the way it is now in most cities, marginally surviving people could be enticed to disperse throughout a city. One of the things that has given Telegraph Avenue and People's Park a bad name in Berkeley is that it has become a kind of mecca for troubled people. Because other cities won't provide a place for them, Berkeley has inherited their problems.

It is also important to remember that many homeless people are capable of helping themselves. A lot of them wouldn't even be out on the street now if it hadn't been for all those urban redevelopment projects that destroyed thousands of cheap hotel rooms and replaced them with shopping malls, boutiques, and high-priced yuppie condos. Many of the homeless are resourceful people. They may not respond to a nine-to-five lifestyle, but will take advantage of less restrictive opportunities. A more active commercial life on sidewalks around city parks will provide incremental opportunities for homeless people to wean themselves from helplessness and despair. Once there are people regularly on the street, selling food and crafts and wares, contact happens. A vendor, who is setting up in the morning or packing up at night, can call out to a street person and offer to pay for a little help. Once some trust is established, that person might watch the stand for a few moments in exchange for something

> *People enjoy and use parks more when something interesting is happening around them.*

to eat and so on. It is important to remember that the street is their "home."

Compared to billion-dollar poverty programs, this incremental approach to the homeless problem may seem hopelessly naive. But it has the virtue of starting with a simple one-on-one level. Until we find a way to create both affordable and convivial housing, street people are here to stay. While some of them might work their way back up the economic ladder and become good burghers, most won't. We might as well ease our guilt, show more compassion and make things more convivial for them. Most people on the streets can't deal with the major bureaucratic institutions of business and government. If they could, they wouldn't be on the streets.

For starters, we should provide simple, inexpensive public bathrooms, street-corner urinals like they have in Europe, and park shower facilities that are functional, airy and safe to use. I'm not talking about building enclosed, concrete bunkers where you can shoot up or get mugged. They could be as simple as a small cement pad surrounded by a metal partition that is open at the top and bottom so as to both air out and discourage loitering. These types of facilities could be built and maintained out of the licensing fees that a renewed sidewalk culture would generate. Making the streets more convivial for bottom-dog people would be more humane and less expensive than the alternative of building, maintaining and staffing ever more care-taking institutions like prisons and mental wards.

I think ecology has something to tell us about the nature of a healthy city. I've asked myself what comparisons might be appropriate between natural ecosystems and

Richard Nilsen

Most of the homeless, like those shown here encamped in Berkeley's People's Park, are resourceful people, scraping a living from the street.

cities. We know that when we clear-cut an uneven-age, deciduous and evergreen climax forest and replace it with a monocultural, even-age species like Douglas fir, we not only change its arboreal characteristics, but also drastically change the surrounding plant and animal community. Similarly, in urban areas, when we bulldoze whole city blocks in the name of urban redevelopment, we destroy the mix of uneven-age buildings and displace their inhabitants in favor of a monoculture that supports a much more limited range of human identity and activity. I've come to see that there is a relationship between maintaining a diversity of architecture and the kinds of people that it attracts.

Though all humans belong to the same species, there is some usefulness in seeing different kinds of people as different kinds of species for the purpose of an intelligent urban planning that will leave all of our lives more interesting, stable and diverse. There need to be affordable places for bohemians, craftspeople, pensioners, sidewalk peddlers, neighborhood junkyards and even ''idlers'' who can become, in Jane Jacobs' phrase, ''the eyes on the street'' that help control crime. The answer to crime is not more cops, but more people with an interest and stake in the street. We can justify this diversity not just on the grounds of tolerance or some kind of noblesse oblige, but even more so out of ecological necessity and a need for liveliness. We know from ecology that those plant and animal communities that are the most diverse are also the most stable and enduring. The presence of diversity is what informs and protects us, it keeps us from becoming stupid. When we have nothing to compare our lives to, we risk following a path that winds up being a dead end.

*GT:* The ecological connection is crucial. We should admit how far-reaching, complex and subtle that connection really is. I believe that if the self-image of a city could assimilate and locate itself within ecological realities, either as fact or ritual, perceptions would change

> *The answer to crime is not more cops, but more people with an interest and stake in the street.*

and new possibilities would emerge.

In 1981, after ten years had passed, I returned to the South Bronx — this time as a designer, paid by a large non-profit organization — to work with community groups putting vacant lots to use. Frankly, I had doubts about how well it was going to go because of the hard times there. But I soon discovered unusual opportunity for urban restoration because of the abundant open space, and the neighborhood vitality that still existed, even with all the violence. Households, city blocks and neighborhoods were still working at scales that were practical and personal.

Today the South Bronx and other inner-city neighborhoods are at pivotal moments in their histories. Large open spaces now exist. Development will occur and construction is imminent. We face a juncture, and people wonder if the failures of modern planning will be repeated or if we can shift toward restoration in the form of urban eco-development.

Let me take you on a tour and contrast two types of development. The first is current normal urban planning. This is what you see driving through the South Bronx — two suburban raised ranch-style houses sitting on a four- to five-acre lot where six-story tenements once stood. There are bars on the windows and garbage is beginning to pile up in the immediate vicinity, where there are burned-out shells of two-family houses. These newer suburban homes could have been picked out of a suburban catalogue. They are referred to as ''living units,'' sharply separated from the land they occupy, from the neighborhood and from each other. Community groups

A good example of mixed-use park space, the Union Square Green Market in New York City encourages people to interact freely with each other, creating a convivial place to congregate. George Tukel in foreground.

Carlos Ortiz

Charlotte Street, South Bronx, 1989. It's easy to look at these ranch-style houses and declare them inappropriate for an urban situation — especially if you happened to grow up in the suburban version. In a city planner's ideal scheme they may be a poor choice, but for the people living in them they are a huge improvement over the decrepit tenements they have replaced.

and city agencies are sponsoring these ''experiments'' in pre-fab modular housing.

Contrast this with a turn toward eco-development. You are sitting with members of a community block association, discussing what local alternatives should be discussed with city officials for the open space surrounding their block. The setting is typical — a three- or four-square-block area, part of which is now open space, part usable housing and commercial buildings, and the remainder abandoned, burned-out buildings. Some observations about eco-development emerge from the discussion:

• The number of people living within the restored area should be at a density that allows for choices rather than unilaterally defining them; in this case medium-density housing interwoven with multi-functional public and green spaces (as opposed to high-rise tenements and concrete retreats and vistas).

• Neighborhood members should be given the opportunity to provide for themselves. Vacant lots should be put to use growing food. Tested and successful solar technologies — rooftop domestic hot-water systems, attached sunspaces and thermosiphoning air panels, for example — should be fully utilized.

• Designs for the neighborhood should incorporate lessons learned from watching healthy natural systems evolve. For example: is the total biomass increasing (more space dedicated to gardens, orchards, parks and green space)? Are material cycles becoming more closed (organic material composted for gardens on a community-wide basis)? Can we have more production from smaller imputs of energy (conservation of energy and its efficient use)?

The whole premise urban restoration as eco-development is based on is that natural systems are easily understandable and make the best place from which to start. People intuitively respond to them, probably because they work in such an economical and elegant manner.

*MH:* Getting involved in neighborhood planning and politics has also shaped my ideas about what a healthy urban ecosystem might look like. Five years ago my family and I moved to the ''M'' Zone in west Berkeley, which historically has allowed for mixed manufacturing, commercial and residential uses. We bought an inexpensive fixer-upper house that was built in the 1920s and watched as real estate values skyrocketed. Simultaneously, we experienced increased pressure from both proponents of urban redevelopment and commercial interests to redefine our neighborhood in ways that would displace the current mix of buildings and people.

One of the things our neighborhood association has fought for, with some success, is to require the city and developers to incorporate older structures, and the people working and living in them, into their plans.

My experience in M Zone politics has also made me question the monocultural zoning policies that characterize most city planning. A tremendous amount of human initiative is thwarted by single-use zoning. It is also expensive and wasteful to separate where we live from where we work. There are millions of self-employed urban people who are forced, by single-use zoning policies, into paying double mortgages and rents in order to run their businesses and maintain their homes. Keeping these two activities isolated from each other is also a major contributor to pollution and traffic congestion. Future urban planning should encourage mixed-use zoning that combines manufacturing, commercial and residential dimensions. We don't need more expensive highways and bridges. They will only foster more pollution, waste more energy and time driving on ''freeways.'' What we need is to encourage live/work communities that cut out commutes and integrate all members of society. ∎

## Shading Our Cities

*If you are an urban resident involved with trees or wanting to be, this is a useful and comprehensive book. Over forty contributors cover everything from keeping city trees healthy to strategies for community involvement with urban forests. Resources like books, periodicals and helpful organizations are also included.*
— *Richard Nilsen*

**Grouping urban trees into blocks rather than rows provides room for root growth, as well as beauty and focus in a courtyard.**

•

One of the earliest and certainly one of the most inspiring greenways is in Denver, Colorado. The city was, like many others, built alongside a river — in this case the South Platte. As the city grew, the river became a dividing line between the rich and poor sides of town, and a sewer for both. So things stood for a century on a river once described as too thick to drink, too thin to plow. Then in 1965 a devastating flood swept through Denver, sending the Platte over its banks and foul water throughout the city. The cost of damage was a third of a billion dollars. Everybody said something should be done. In 1973, another flood hit. An engineering report was prepared: the price tag for fixing up the river would be $630 million.

At that point, Denver had $1.9 million to use on the river. What to do? The answer was to call in Joe Shoemaker, hard-nosed Republican state legislator, a lawyer and ex-navy man, who, as he says of himself, doesn't take any crap from anybody. His idea was, instead of trying to pass a $630 million bond for flood control, which would probably be impossible anyway, why not set up a foundation and "return the river to the people" — not as a rip-rapped channel, but as a greenway.

And so it is today. Joe Shoemaker and a

### Shading Our Cities

Gary Moll and Sara Ebenreck, Editors
1989; 333 pp.

**$19.95** ($21.95 postpaid) from Island Press, Star Route 1/Box 7, Covelo, CA 95428; 800/828-1302 (or Whole Earth Access).

cadre of devoted young colleagues (now including his own son, also a state legislator) created the Platte River Greenway from one end of the city to the other (and nowadays, beyond) by assembling a foundation board of directors made up of powerful figures who might have been enemies, by resolutely avoiding the accession of any government trappings whatever ("to have no power is to have all power," says Shoemaker), and by piecing together from scores of public and private sources some $15 million to do the job. The result is a greenway that provides 450 acres of riverside parks, forty miles of interconnected hiking and biking routes, and a river that brings the city together rather than dividing it.

## A Green City Program

*Urban sustainability is no ecological sideshow. In fact, with 75 percent of the population of the United States living in urban areas, it's fair to say it's the main attraction. This book skips the Cassandra-on-the-tightrope portion of the show (in which we learn about the ecological crimes of industrial society) and moves right into the grand finale: hope for the future based on practical sustainable living practices — a Green City program.*

*Green City was generated from the proceedings of a series of San Francisco Bay Area meetings on moving the region toward self-reliance, but it serves as a design model for any urban area. The diversity and experience of the participants — who ranged from governmental agency officials to the director of Golden Gate Park to a vast assortment of environmental and community activists — is nicely reflected in the accessible tone and useful structure of the book. Each section (on urban wild habitat, diversified transportation, renewable energy, small business, and more) outlines both short- and long-term action proposals and includes a visionary fable of life in the new "Green City." Anyone interested in the flourishing of cities — backyard gardeners, city council members, community activists, or high-school students, for example — might want a copy of this book. Green cities could turn out to be the greatest show on earth.*
— *Jeanne Carstensen*

•

*Recycling and reuse in Green City: what's possible?*

In Green City, the crude garbage cans of the 1980s are a thing of the past. In their place are sophisticated containers, each with several compartments: for tin, glass, cardboard, newspaper, a couple of grades of paper and miscellany. Aluminum cans have practically vanished, in the wake of deposit legislation that made it more economical to supply beverages in refillable bottles; most glass is reused because of those laws. The primitive garbage trucks of the past have been replaced by more elegant models with multiple compartments to match those on the household recycling containers. Household kitchen scraps are set out separately and picked up by a fellow who lives on the block. He collects

### A Green City Program

Peter Berg, Beryl Magilavy, and Seth Zuckerman. 1989; 70 pp.

**$7** postpaid from Planet Drum Foundation, P. O. Box 31251, San Francisco, CA 94131; 415/285-6556 (or Whole Earth Access).

the stuff in a hand-drawn wagon. Some of it he feeds to his chickens; the rest he composts (along with the chicken manure) in bins in the community garden. For these services he draws a small stipend from the city, which pays him out of the disposal fees from the block. The neighbors then use the compost on their vegetable plots, and enjoy the benefit of the rich humus without the task of shoveling their kitchen scraps in with straw and soil.

## Global Releaf

*The venerable American Forestry Association has caught up with the times with a national tree-planting program tied to the threat of global warming and called Global Releaf. They have several handbooks available, all designed to assist community tree-planting groups. One of their goals is to help get one million new trees growing on private land in U.S. cities by 1992.* — *Richard Nilsen*

### Global Releaf

Information **free** from The American Forestry Association, P. O. Box 2000, Washington, DC 20013; 202/667-3300.

•

Among the titles offered:

*Tree Care Handbook*
*Save Our Urban Trees*
*Gypsy Moth Handbook*
*Forest Effects of Air Pollution*
*Global Releaf Brochures*

# Urban Designs That Work

## BY DAVID SUCHER

Paul F. Housel

The First Interstate Plaza in Seattle. This well-used street-level plaza includes pedestrian circulation, seating, plants, and a sunny exposure. In good weather, the plaza is used for a variety of events. This plaza was designed well from the start, but it can stand as inspiration — the urban world is full of empty, wind-swept plazas and sidewalks (many orphans of the sixties) waiting to be filled up with uses which can create a richer and more human environment.

*David Sucher is a Seattle real estate developer who is helping his city discover that in order to change the urban scene, the place to start is with the urban mind-set; and that for doing either one, praise works better than criticism.*
*—Richard Nilsen*

O VER the last couple of years, I've helped to establish an urban planning process in Seattle called the Urban Designs That Work Awards. Now in its third year, the program spotlights those buildings/spaces or parts of buildings/spaces which ''work'' and are worthy of recognition and emulation. The Seattle Design Commission — a public body — oversees the selection of successful projects and then hosts a public forum to present awards to project developers and architects.

A key element here is public discussion of what makes the project work. Usually, an architectural award consists of a plaque and a picture in a magazine, with no discussion of what is going on in the building. Traditional awards function more as marketing tools for architects and builders than as learning tools for the public.

My hope is that Urban Designs That Work will become an environmental education process. We need to become aware of our built environment, understand what is good about it and repeat those details which are successful. My own long-term goal for this process is to create a positive feedback loop that encourages successful urban design.

An architect once said: ''There are truths about floorplans.'' There are also truths about cityscapes. We are all really copycats at heart. It's easier and safer that way. Why should we reinvent the wheel with every effort? In our daily lives we learn to sort out what works from what doesn't work. We learn to repeat successful behavior and discard the failures. Unfortunately there is an ongoing myth of the

Paul Hernandez, Seattle Design Commission

The espresso bar in front of Nordstrom's clothing store. What had been an ''empty space'' under a marquee is now a place to relax, with espresso cart, tables, and chairs. Seattle's average of 162 rainy days per year make overhead cover essential to relaxation.

''architect as artist,'' fighting against bourgeois conventionality. The problem with this attitude is that it puts too much emphasis on self-expression and novelty rather than on designs that work well.

Great cities come not from great buildings but from great zoning codes. Great work can come from tight limits as easily as from freedom, and is often a variation on a theme. ''Design flexibility'' and ''innovative design'' can be the enemies of urban architecture because they encourage builders and designers to do something new just so that they can attract professional attention. A truly innovative approach would be to search through the relevant environment, pick out specific buildings that were successful by community standards, and repeat them. But the prevailing image of the architect as artist — per Ayn Rand — interferes.

One of the hot political issues Seattle has struggled with over the last few years is how to design multi-family (apartment and condominium) structures. In the attempt to hammer out regulations, the City assembled a group of developers and neighborhood activists to act as a sounding board, and as a real estate developer, I became involved. We must have

met 20 times to go over draft language on everything from the protection of wildlife to shadow impacts.

The neighborhood people kept talking about ''excessive height, bulk and scale.'' I wondered what that really meant. Or rather, I wondered what kind of structures would satisfy them and I realized that in none of our discussions did we ever speak of specific buildings, either pro or con. It was all very abstract. No one ever said: ''That new building on XYZ Street in my neighborhood is truly horrible'' or ''I wouldn't mind living near a building like the one at the corner of A and B Avenues.'' Through all those meetings, we never created a visual frame of reference, never had models to which we could refer.

I was never sure if the anti-apartment feelings were a principled reaction to ugliness/inappropriate design, or an unprincipled desire to keep possibly lower-income renters away from single-family neighborhoods. I wondered if there could be any apartment building which would be good enough to please the neighbors. Since we never spoke in terms of specific buildings, it was impossible to know.

As I sat there, I realized how useful it would be if we could talk about

buildings and details of buildings that we liked or didn't like. It would first be necessary to recognize these differences, and my Awards scheme seemed like a way to do this. Since I had worked for the Seattle city government earlier in my life and had also dealt with the bureaucrats on a regular basis because I develop real estate, I approached individuals whom I thought might be receptive. There was polite interest and sympathy, but no action.

Soon thereafter a local architects' group, Arcade & Blueprint, held its annual show — Call For New Work — at which 2' x 3' panels displaying new projects were on display. I made up a panel outlining my ideas for the Awards. To add symbolic significance and visual interest, I attached a large county-fair blue ribbon to the poster. It was accepted and hung and prompted some discussion at the event, but again no action.

A couple of months later, I produced a miniature version of the poster and sent one to the Mayor and each of the nine council-persons. I sent posters to the press too, and one local columnist thought it was a great notion and devoted a glowing column to it. That helped a lot because it gave external validation to the idea. Simultaneously, I went to talk to a friend at the Seattle Design Commission. She liked the idea, talked it up to the rest of her group and they asked for $500 from the City Council to pay for the costs of the event.

The Design Commission selected five sites in downtown Seattle which were worthy of recognition. An out-of-town speaker was invited (they are always smarter if they're from out-of-town, you know) and the first awards were given out in August, 1988. We got good press, the public responded favorably, and the politicians were delighted.

In the long run, The Urban Designs That Work Awards should become an ongoing civic process to spotlight what people like and to then feed that back into the zoning code. It should work within the guts of the building regulation process to develop common frames of reference using real buildings. When a developer applies for a permit, both the City and the neighborhood could engage in discussion with reference to real buildings: ''that one we like''; ''that one was not quite so popular.'' Right now developers are working in a vacuum — the land-use code says one thing, but the City also has ''environmental policy'' hoops to jump through, and those are the real regulations. It's very confusing and of

no benefit to anyone, because there are no clear models.

As an aside, I've thought it curious that one could not build an ''Anhalt'' apartment building in Seattle today. Fred Anhalt was a builder of very lovely Seattle apartments in the late twenties and thirties. They are still very popular, and command good rents. They fit in very well with their neighborhoods. Yet you couldn't build one today because many of them come right up to the sidewalk, and that's not allowed. Even though these buildings ''work,'' some of their most critical elements violate the current rules.

Perhaps we ought to have a process in which successful building types are pre-approved. You buy a 60' x 120' lot, and you go downtown, and they've got a catalog of basic design concepts which would be acceptable for that type of lot in that neighborhood. By starting with one of these designs, developers have some security that their proposals will be approved, and the public has some guarantee it won't have to do battle against another eyesore. This would not be cheap, of course, but then neither is the current appeals process, or lawsuits after the fact. These designs would be Seattle ''standards'' (for Seattle only, of course) to which we could refer. It would be a visual

vocabulary — perhaps a shorthand for expressing what's considered good behavior in a building.

The gist of the Urban Designs That Work is that now the public, the building professions and the government have a way to recognize good things about buildings and to praise and explain why they are good — a way to spotlight those elements of our Seattle environment which are worthy of applause and emulation.

One of the local chapters of the AIA (American Institute of Architects) has an ''Orchids & Onions'' program; the orchids are fine, but it's counterproductive to point out in public where someone has made a mistake if you want to influence their future behavior. I believe that we should accentuate the positive. For the foreseeable future, the vast majority of decisions about the shape of the built environment will be made by private parties with only nominal regulation.

Building is an emotional process. It's also a very public process. It takes nerve and ego to put oneself on the public display inherent in putting up a structure. It makes no sense to hold up the ''onions'' to public ridicule because the long term strategy must be to sensitize builders, not to close off their hearts to designs which contribute to their communities. ■

Fred Anhalt was a Seattle builder in the 1920s, when setbacks from the property line were not required. Because of current code restrictions, this handsome Anhalt apartment building could not be built in Seattle today, because the walls, visible on the left side of the photo, come right up to the sidewalk.

David Sucher

## Society for Ecological Restoration

*This organization serves a new profession — environmental restorationists — although membership in the Society is open to anyone interested in the subject. One of the healthy things about restoration work is that it involves diverse professional specialties in a generalist pursuit. This is also part of the challenge for a new discipline, since anyone can call himself a practitioner of environmental restoration. SER is where the philosophy, politics and business of restoration are getting thrashed out, including issues like setting standards for gauging the relative success of restoration projects.*
*—Richard Nilsen*

**For membership information for the Society for Ecological Restoration**, write to the University of Wisconsin Arboretum (address at right).

## Restoration & Management Notes

*This "bulletin board" of environmental restoration does a nice job of serving both a profession and a movement. It's the best single source for restoration news. Because this is a new field full of pioneering spirit, the writing is usually straightforward, and the stultifying peer-reviewed prose found in so many scientific journals is missing. There are long articles, plus abstracts from diverse small publications and conference proceedings (see p. 130 for some of what is routinely scanned for relevant news). To the casual observer two things are apparent — there is a lot of restoration work going on, and it is beginning to happen all over the world.* —Richard Nilsen

### Restoration & Management Notes
William R. Jordan III, Editor
**$15**/year (2 issues); **free** with membership in Society for Ecological Restoration ($30) from University of Wisconsin Arboretum, 1207 Seminole Highway, Madison, WI 53711; 608/263-7888.

•

What the restorationist creates is rarely, if ever, a finished product. Rather it is a beginning — often an early stage in succession. And if the restoration is to mean anything at all in the long run, attention must be paid to its likely course of development in the future. Thus, the question becomes not just what exists at any particular time, but what processes have been set in motion, what dynamics prevail? In short, how successfully has the project been *aimed*?

•

Let us be clear that we are unabashed fans of Disneyland. In our opinion, Disneyland is one of the finest things done for people by people. Among other things, it creates tangible fantasy and apparent reality in ways that are pleasing to most of its visitors. But it is not reality. Let us further be clear that the fantasy of a "Disneyland" is better than the reality of another suburban parking lot. Similarly, if a truly native ecosystem cannot be restored, then restoration of something biologically viable and sustainable is far preferable to the complete loss of that ecosystem.

Exactly what are the factors that distinguish a "Disneyland" from a restored native ecosystem? One set of factors that has so far been treated only superficially is the genetics of restoration. The genetic nature of introduced stock can profoundly influence the behavior of the individuals, which in turn may affect the dynamics of the entire community and disrupt or alter the course of co-evolution within the community. All of these effects are of great concern to the restorationist.

**Retiree volunteers from the Chevron Corporation pair off to prepare holes and plant valley oak seed at The Nature Conservancy's Cosumnes River, California, site.**

**Using the Bible as a guide, restorationists in Israel are creating a landscape that evokes the landscape of the prophets.**

# 3.

# ENVIRONMENTAL RESTORATION AROUND THE WORLD

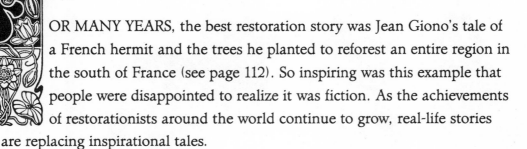

OR MANY YEARS, the best restoration story was Jean Giono's tale of a French hermit and the trees he planted to reforest an entire region in the south of France (see page 112). So inspiring was this example that people were disappointed to realize it was fiction. As the achievements of restorationists around the world continue to grow, real-life stories are replacing inspirational tales.

In the category of restoration due to individual perseverance, David Wingate's work providing bird habitat on Nonesuch Island in Bermuda is hard to match. Part of what makes this story so compelling is that it presents environmental restoration stripped to the bare essentials — one rare bird species, one tiny, deserted and degraded island, and one person's life work. Islands afford such splendid isolation, but restoration elsewhere is rarely so simple. It's tempting to say that people get in the way, but that's a distortion, and it can subvert the long-term effects of this work. Environmental restoration is not about building better fences: people are always part of the equation. Here are Indians facing up to the effects of overgrazing, Australians coping with salted soils produced by modern agriculture, and Russians grappling with horrendous water pollution. We face the same challenge around the world: to make people aware that their actions have environmental consequences, and to provide incentives for them to change. Redirecting human behavior is more difficult than the physical acts of restoration.

# THE RESTORATION OF AN ISLAND ECOLOGY

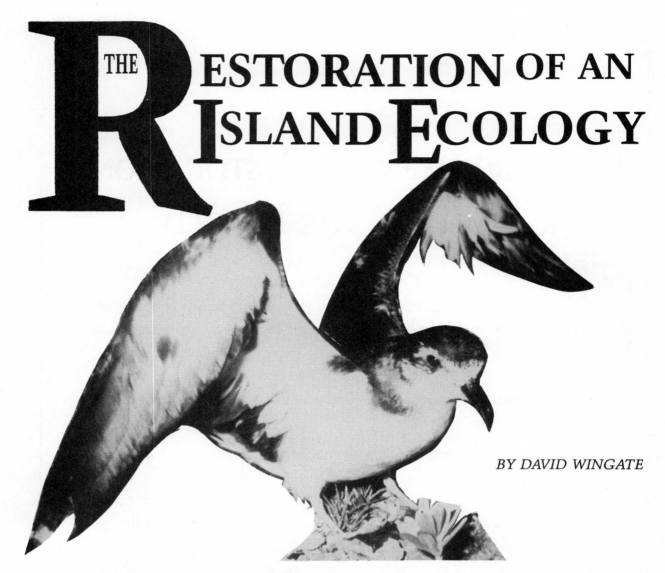

*BY DAVID WINGATE*

**The true story of the man who planted eight thousand trees and resurrected an "extinct" bird.**

IN 1951, THE WORLD SCIENTIFIC COMMUNITY was stunned by the announcement that the cahow — a bird whose name had become synonymous with extinction because it was thought to have become extinct in the 1600s, around the same time as the dodo — had just been rediscovered. The cahow is a member of the petrel family, in the order which contains albatrosses and shearwaters. It ranges widely in the North Atlantic to the western edge of the Gulf Stream, where it feeds on squid and fish, but it breeds only on the 20 square miles of oceanic islands of Bermuda, located at 32° N and 64° W in the western reaches of the Sargasso Sea, 580 miles east of Cape Hatteras.

As the conservation program launched to save this extraordinary bird from extinction has gradually succeeded against all odds, expanding into the restoration of an entire terrestrial ecosystem on 15-acre Nonsuch Island, the name "cahow" has ultimately attained wider significance as a symbol of hope for conservationists around the world.

*David Wingate has spent his adult life nursing one species of pelagic seabird back from the edge of extinction. He set out to save a bird — the cahow — but in the process he has restored the ecosystem of an entire island. With nature it is impossible to do just one thing. Like many a beautiful place, Bermuda has become a victim of its own success. Wingate serves as the island's Conservation Officer. If you would like to help fund an endowment to guarantee the cahow's future, contributions may be sent to: Cahow Account, Bermuda Audubon Society, P. O. Box HM 1328, Hamilton, Bermuda.*
*—Richard Nilsen*

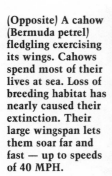

(Opposite) A cahow (Bermuda petrel) fledgling exercising its wings. Cahows spend most of their lives at sea. Loss of breeding habitat has nearly caused their extinction. Their large wingspan lets them soar far and fast — up to speeds of 40 MPH.

(Right and below) Nonsuch Island, Bermuda.

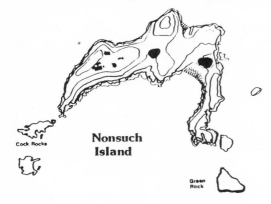

Cock Rocks

**Nonsuch Island**

Green Rock

The history of man on Bermuda provides a stark contrast to the story of the cahow. Settled as a British colony on the strategic sea lanes between the old and new world, the island has become so successful economically that it is now threatened with environmental self-destruction. Bermuda is now the most densely populated, isolated geographic and political unit in the world, with a density of five people (and two houses) per acre and a growth rate of more than 500 new housing units a year.

The problems that conservationists confront in trying to resurrect the cahow, and the fragile oceanic island ecosystem that it symbolizes, can only be appreciated within the broader context of this human history. In telling the story of man and the cahow on Bermuda together, from pre-colonial time until the present, I want to try to convey in a chronological perspective what it is like to be involved with a very long-time restoration project — the patience required, the drudgery, the occasional agonizing setback, and, finally, those exhilarating breakthroughs that make it all seem worthwhile. Only in this way does it become apparent how closely the fate of these two species has become linked.

OUR STORY BEGINS more than 400 years ago when Bermuda was first discovered by Portuguese and Spanish navigators exploring the New World. In those days the treasure-laden galleons from the Spanish Main used to sail north from the West Indies to catch the westerly winds for their return home. Many came to grief in sudden, violent storms on Bermuda's uncharted reefs. As darkness overtook the stranded survivors they were terrified by the hordes of nocturnal seabirds coming and going to and from their nesting grounds each night. The sailors took them for evil spirits and named Bermuda "The Isle of Devils". The Spanish never settled Bermuda, but they left a legacy of wild hogs behind to provide food for future shipwrecked mariners. The hogs caused such untold havoc among the seabirds that they ultimately destroyed far more than they provided.

It was in circumstances similar to those of the Spanish that the British first landed on Bermuda. In 1609, a fleet sailing to relieve the Virginia Colony was dispersed by a hurricane near Bermuda and the flagship, the *Sea Venture*, was shipwrecked on its shores. The survivors set about building ships to make their escape. It took them nine months, and in that period Sir George Somers became so impressed by the island's natural beauty and virgin resources that he determined to start a colony.

In the clear surrounding waters the fish were so tame they could be caught by hand. The land itself was covered in dense forest, and two trees in particular were especially common. The Bermuda cedar provided valuable timber for ships and the palmetto provided leaves for thatching the huts and making ropes and basketware. Both trees provided edible berries for food.

But apart from the pigs released by the Spanish it was a land devoid of mammals. Indeed, the only four-footed creature to reach Bermuda before man arrived was a small lizard of the skink family. An abundance of sea turtles hauled themselves up on the beaches to bask in the sun or lay their eggs. But by far the most dominant element of the fauna was the birds, because these had no difficulties in colonizing the island across the ocean.

There were landbirds of several species, so tame that they readily landed on the settlers' shoulders. We do not know all the species involved because many were soon to be exterminated by the impact of human settlement. Seabirds were even more abundant, because they were adapted to exploiting the food supply from a vast area of surrounding ocean. By day, tropicbirds or longtails, as we have come to know them, were conspicuous.

These diurnally active seabirds were eclipsed at night by nocturnally active shearwaters and petrels in even larger numbers. One of these, which came to be known as the cahow, outnumbered all of the others put together. The cahow was a ground-nesting, soil-burrowing seabird, and it nested both along the coast and inland, under the forest canopy. Cahows are also some of the fastest and most efficient flyers in the world, and it was this extraordinary ability that enabled them to reach beyond the relatively sterile waters of the Sargasso Sea to feed in the rich upwellings of the Gulf Stream more than 400 miles away.

This then was the island that Sir George Somers and his party had stumbled upon. Although Sir George himself never lived to see it, his dream was realized three years later with the colonization of Bermuda by a group of wealthy investors, who formed the Bermuda Company.

BY 1618 THE ISLAND WAS SURVEYED AND SUBdivided into narrow strip shares. There were no roads and the idea was to give each parcel access to the sea. The main island of Bermuda is narrow — only one mile across at its widest point. But the strips ran at right angles to the lay of the land, and this arbitrary surveyor's solution has caused problems to this day.

Colonizers were invited to settle as sharecroppers, but defending the island against a feared invasion by the Spanish took so much time and energy that the settlers were forced to live mainly off the virgin resources. These proved to be so ecologically unstable and fragile that they were soon exhausted.

As on other oceanic islands, the fauna of Bermuda proved exceptionally vulnerable because it had evolved in the absence of man and other mammal predators and showed no fear or defences against them when they eventually arrived.

Legislation was passed in order to save the sea turtles as early as 1620. Seabirds were destroyed not only by man himself, but by a plague of rats which reached the islands accidentally in 1614 and by the cats and

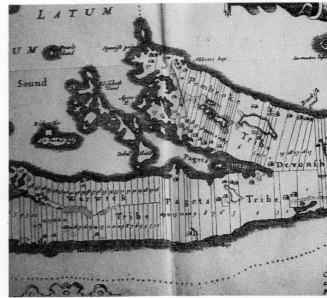

Norwood Surv

dogs which were brought in to control the rats. In less than thirty years the abundant cahow was reduced to the verge of extinction and only its fossil bones — which remain in caves and manmade excavations even today — attest to its former abundance.

The virgin forest fared no better as the settlers set to work with axes to fell timber to be shipped back to England, or to build ships and houses, or simply to clear the land for such food and cash crops as corn and tobacco. As a remedy, at the height of the rat plague in 1616, Governor Tucker ordered a general burning of the island, which laid waste to large areas of land.

By 1684 it was all over. The Bermuda Company, having exhausted the virgin resources, disbanded and those settlers who remained were ultimately forced to go it alone, and to learn a new way of life in order to live in balance with the land. Thus evolved a race of true Bermudians who disdained to work the land, but like the cahows turned their attention back to the sea as a source of survival. Fishing the Banks, hunting whales and trading far and wide were carried on in ships built of Bermuda cedar, which fortuitously was the one native tree which seemed to thrive on man's abuse of the land until it soon became completely dominant in the flora.

The more successful of the ships' captains built sturdy houses of Bermuda limestone which, with their whitewashed stone roofs designed to channel rainwater into cisterns, were themselves a masterpiece of cultural evolution and adaptation. (Bermuda has brackish marshes with little open freshwater, but with 60 inches of annual rainfall, the roof collectors and cisterns not only made human habitation possible, they allowed it on any part of the island.) As commerce with the outside world continued, there was a selective introduction of plants either useful or

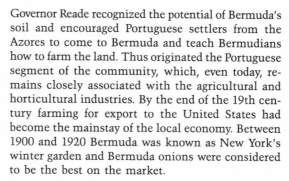

Surveyors love straight lines, and in this 1618 survey they were put to seemingly good use. On roadless colonial Bermuda, the narrow parcels guaranteed each colonist access to the sea. But this arbitrary solution totally ignored the lay of the land and has been a problem ever since.

On an island with scarce fresh groundwater but lots of rain, this is the human adaptation that evolved and is still used throughout Bermuda — a whitewashed limestone house with stone roof to collect rain into a cistern.

Governor Reade recognized the potential of Bermuda's soil and encouraged Portuguese settlers from the Azores to come to Bermuda and teach Bermudians how to farm the land. Thus originated the Portuguese segment of the community, which, even today, remains closely associated with the agricultural and horticultural industries. By the end of the 19th century farming for export to the United States had become the mainstay of the local economy. Between 1900 and 1920 Bermuda was known as New York's winter garden and Bermuda onions were considered to be the best on the market.

The third significant development was the sudden growth of Bermuda's strategic significance, following the Revolutionary War and the Declaration of Independence by the American colonies. Bermuda was now the only base for British naval and military operations in the New World between Halifax and the West Indies. As the Empire reached its zenith in the Victorian era, millions of pounds sterling were poured into the island for defense purposes and it became known as the "Gibraltar of the West". A whole chain of massive new fortifications sprang up around the periphery of the island.

As the population grew, the little white-roofed houses began dotting the landscape in ever-increasing numbers, but it was a population now dependent on, and largely sustained by, strategic and economic factors in the outside world.

The middle and late 19th century was a time of great cultural infusion and contact between Bermuda and the outside world. Many of the officers of the garrisons had an interest in the natural sciences and their explorations provided the first detailed scientific descriptions of Bermuda's geology and natural history, beginning about 1840. Among other things their curiosity inspired them to look into the almost legendary accounts of the cahow bird by the early settlers. Their investigations were concentrated on the remote Castle Harbour Islands, where local fishermen continued to report nocturnally active seabirds which were indiscriminately referred to as pimlico or cahows. These islands were investigated by a number of naturalists between 1840 and 1900, but the only nocturnal seabird they ever succeeded in finding was the pimlico, or Audubon's shearwater. This led some to conclude that "cahow" was a synonym for the shearwater, but others, like the famous naturalist Addison E. Verrill, concluded that the cahow must have been a species of auk!

The confusion lingered on into the early 1900s, when

beautiful, such as citrus, bananas, hibiscus, oleander and other exotic trees and shrubs. Many of these went wild and gave new character and colour to Bermuda's landscape.

For nearly two centuries, Bermudians built ships of native cedar and roamed the world as traders, privateers and even in some instances as outright pirates. It was a tough life, during which Bermuda was never able to sustain a population of more than eight or ten thousand people.

WITH THE BEGINNING OF THE 19th CENTURY, three developments enabled the population to begin a gradual increase. The first was the industrial revolution and the general improvement in medical services and health care. The second was a revitalization of Bermuda's agricultural industry. In the mid-1800s new evidence clarifying the identity of the bird was discovered in the form of abundant fossil bone deposits in Bermuda limestone caves. By 1915 these bones had been examined by the avian paleontologist R. W. Shufeldt at Carnegie Museum, who identified them as a species of gadfly petrel, quite distinct from the bones of the Audubon's shearwater which also occurred, though in lesser numbers, in the same caves. It was the sheer abundance of these bones together with the pronounced hooked bill that led Shufeldt to conclude that they must represent the legendary hook-billed cahows of the early settlement days.

Then an amazing fact came to light. In 1906 Louis Mowbray, a Bermudian naturalist, had actually collected a living gadfly petrel from a crevice on one of the Castle Harbour Islands. The specimen had been preserved and sent to the American Museum of Natural History. Its bones were now compared with the fossils and found to be identical! Thus was the type specimen of the Bermuda petrel — *Pterodroma cahow* — described in 1916.

It may seem incredible now that no one followed up on this revelation at the time, until we remember that this was during the Great War, when an entire generation of European and American men were wiped out in the muddy trenches of Europe. It was also before the age of conservation when the emphasis was still on collecting and cataloging of species in the world's museums, rather than studying the living organisms in their actual environment.

Thirty years were to pass before another specimen of the cahow turned up: this time a fledgling which was killed when it flew against St. Davids lighthouse in June 1935. The specimen was taken to Dr. William Beebe, who was based on Bermuda doing fish studies with the New York Zoological Society at the time. Beebe was an ardent naturalist and prolific writer who was one of the first to popularize natural history subjects for the general public. He would have liked nothing better than to claim the rediscovery of a bird that had been presumed extinct. Nevertheless, he continued to be confused between the cahow and the shearwater and his photographs confirm that the shearwater was once again the only species whose nests were found. It was Dr. Robert Cushman Murphy, Curator of Birds at the American Museum of Natural History and a world authority on seabirds, who eventually provided the identification of this second known cahow specimen.

INCREASED CONTACT WITH THE OUTSIDE world in the Victorian era also greatly accelerated the number of new plant and animal introductions — a process that was often officially encouraged by the governors, such as John Lefroy. In the 1870s, the beautiful European goldfinch became established as a cage-bird escapee. The giant toads and diminutive noisy whistling frogs were deliberately introduced at about the same time. Even the pretty anolis lizard, which

we mistakenly call a chameleon, and think of as a true Bermudian, is a fairly recent arrival, deliberately introduced from Jamaica in 1905.

Bermuda continued to evolve into a kind of semitropical paradise. It was this circumstance, along with a quaint and easy-going way of life — still dependent mainly on bicycles and horses for transport long after the motor car was introduced on the continents — that led to the development of yet another industry, tourism. This new industry soon became the dominant source of revenue for the island.

Wealthy families were able to hop onto a clipper ship in New York and spend their winters on Bermuda, away from the cold and snow. As steamships began to replace sail, this progress was accelerated. A number of large hotels and cottage colonies sprang up and by the 1930s Bermuda was a bustling community with a railway running down the length of the island.

Agriculture was now in decline due to unfavourable tariffs imposed by the United States. However, the farming industry received one final boost after the outbreak of war between America and Japan when the trade in Japanese Easter lilies was suddenly cut off. Bermuda was able to supply the American market with Easter lilies of its own.

When World War II broke out in 1939 Bermuda's strategic location was once again to play an important role. As a result of a wartime lease arrangement between the U.S. and Great Britain, a large area on the western side of St. Davids Island was bulldozed into the sea and mixed with dredged sediment from Castle Harbour to form an airport. Cooper's Island was now linked by dredged fill to St. Davids Island and transformed into a complex of water catchments and underground fuel and ammunition storage bunkers. (Since then, it has even entered the space age with the building of the first down-range NASA tracking station.) Soon aircraft were roaring where only mangrove swamps and quiet backwaters had been before.

It was a traumatic time for the St. Davids Islanders, but from the perspective of the rest of Bermuda, this gargantuan ecological transformation was perceived as having many potential benefits.

As late as 1945 Bermuda retained its rural aspect with dense cedar forests crowning the hills and Easter lilies growing in the valleys. But Bermuda was soon to face another ecological catastrophe.

THE ADVENT OF AN AIRPORT HAD GREATLY facilitated the process of new plant and animal introductions, especially those species (such as insects) with a short life span, which had previously had difficulty surviving the transit time to the islands. In the mid-1940s two small scale insects were accidentally imported to Bermuda on ornamental junipers from California. The Bermuda cedar (actually a juniper) had evolved in isolation from such pests and had no natural resistance to these scales. There were no

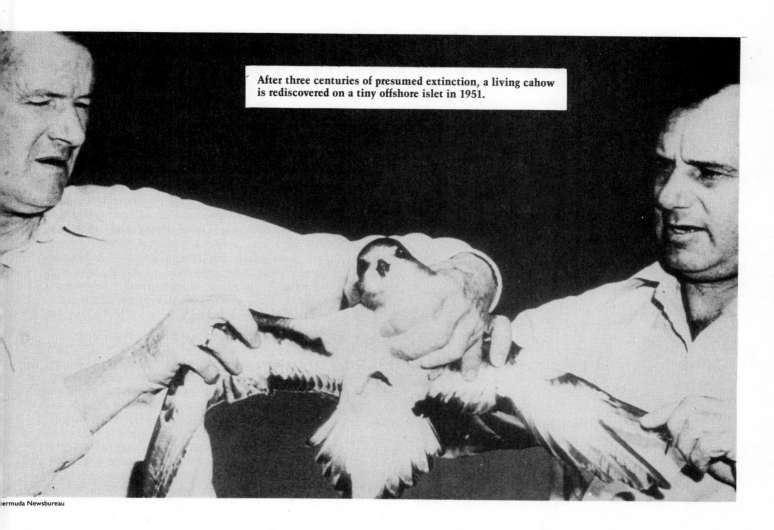

After three centuries of presumed extinction, a living cahow is rediscovered on a tiny offshore islet in 1951.

ermuda Newsbureau

parasites or predators to keep them under control. In the space of only three years, between 1948 and 1951, more than 96 percent of those trees which had formed a virtual monoculture on the island, were destroyed. It was a harsh reminder, once again, of the ecological fragility that is so characteristic of oceanic island fauna and flora.

Due to the cedar forest decimation, for the first time in their history Bermudians began to think seriously about the need for conservation. No event could have been more timely to focus this concern, or to give hope that all need not be lost, than the rediscovery in 1951 of the cahow by Dr. Robert Cushman Murphy and Louis S. Mowbray. (The latter was the son of the man who collected the original specimen; he had also become curator of the Bermuda Government Aquarium, in the footsteps of his father.)

It seemed inconceivable that the cahow, the most vulnerable species of all, which had declined to oblivion within the first few years of Bermuda's settlement, could possibly have survived up to and beyond such recent traumatic events as the bulldozing of the airport — which destroyed more than half of its original nesting islands — and the loss of the cedar. Nevertheless, a third cahow was found washed up dead on the beach of Cooper's Island in 1945 and provided the inspiration for a last-ditch search expedition.

I was only a schoolboy at the time, but my budding interest in birds secured me an invitation to join the expedition on the day of rediscovery and I will never forget the elation on Dr. Murphy's face when he and Mowbray succeeded in noosing a bird out of its deep nesting crevice, held it up to the light, and exclaimed, ''By Gad, the cahow!''

Incredibly it had survived for over three and a half centuries on a few inconsequential offshore islets, totalling no more than three acres in area, which the mammal predators had been unable to colonize. The government immediately declared these islands Sanctuaries in 1951 and a conservation programme was launched to try to help the bird.

⌁

TO REPLACE THE DEAD CEDAR FOREST, A MASsive reforestation programme was begun at about the same time. The emphasis was on fast-growing exotic species like the Australian casuarina tree and a variety of more colourful ornamentals such as the poinciana. Concentrated on roadsides and government-owned lands at first, it eventually led to the establishment of parks and formal gardens for the recreational benefit of the general public. Reserves for the conservation of native flora and fauna generally were still largely neglected — as illustrated by the government's policy towards garbage disposal. Indeed, it was the use of Bermuda's marshlands for garbage disposal, with the intent of eliminating them in pursuit of a "final solution" to the mosquito control problem, that led eventually to the formation of such non-governmental conservation agencies as the Bermuda National Trust and the Bermuda Audubon Society.

The opening of the new airport to commercial air traffic after the war brought a rapid expansion of the tourist trade. In the past, our tourism had been largely restricted to smaller hotels and cottage colonies, but the explosive demand for more rooms ultimately led to multi-storey, convention-oriented hotels.

Such massive constructions posed new problems for water provision, sewage treatment and especially for transportation. The introduction of the motor car after 1946, to narrow and winding roads, gave rise to immediate problems which have since grown to nightmarish proportions as a rapidly increasing material standard of living enabled nearly everyone to afford cars.

The decade of the 1960s saw such a rape of the environment for new real estate subdivisions that the government was forced to act by imposing stringent new planning regulations. But even though a plan-

ning department was established in 1965 and a succession of development plans with different categories of zoning have been set into law, the fact remains that between 300 and 600 new housing units are still being added to the diminutive and finite landscape each year. Farmland, once the mainstay of our economy, has dwindled from 3,000 acres in 1920 to a low of 320 acres by 1987.

The most insidious effect of urbanization in Bermuda has been the steady impoverishment of our natural environment. Now that the housing densities have risen to more than two per acre, the intervening open spaces have become so fragmented and degraded by wind exposure that they are often deemed no longer worthy of maintaining in a natural state. The most frightening aspect of this trend is that it involves the entire island and is wreaking havoc with that resource of natural beauty, peace and tranquility — the Bermuda Image — which is the fundamental selling point of our tourist-dominated economy.

Bermuda is turning into a city without a countryside. Open spaces survive only as a few isolated green islands, completely surrounded by suburbia. We are about to reach a stage where the only open space to remain will be that which has been deliberately set aside in a national system of recreational areas, parks and nature reserves. Work is already well under way on the establishment of a National Parks System aided by a National Parks Act passed in 1986, but to date only 600 acres of land have been secured for passive recreational use and environmental conservation. This is a costly process, because back in 1618 the *entire* island had been subdivided. All land needed subsequently for schools, parks, military forts — to say nothing of ecological preserves — has had to be bought back from private owners. To give some idea of the small scale and intensity of land use on Bermuda, the largest acreage of open space so far set aside is dedicated to intensive recreation, and 580 acres of that, or 5 percent of Bermuda's land area, is devoted to golf courses.

The task at hand is urgent and complex but the miracle of Bermuda is that despite such intense development and abuse so many fragile features of our unique natural heritage survive in small pockets here and there. We haven't lost a species since 1900.

**Pembroke Parish, modern-day Bermuda. With an average housing density on the island in excess of two houses per acre, maintaining habitat for native plants and animals is increasingly difficult.**

OF ALL THE NATURE RESERVES WHICH HAVE been established on Bermuda since the 1960s none has attracted more attention internationally than the Castle Harbour Island group where the rediscovery of the cahow in 1951 has inspired an ambitious restoration effort to create a Living Museum of precolonial Bermuda on 15-acre Nonsuch Island.

Nonsuch Island differs from most other nature reserves in that it is a product of *restoration* rather than protection. When first established in 1961 it was essentially a desert island, having formerly been used

In 1964, sixteen years after the widespread destruction of cedar trees on Bermuda by imported scale insects, restoration of Nonsuch Island begins with small grass-high plantings of native palmettos amidst the dead cedars.

as a yellow fever quarantine hospital and later as a reformatory. Its once-magnificent cedar forest had been entirely destroyed by the cedar scale epidemic; wind, salt spray and free-roaming goats had reduced the remaining vegetation to a dense grass cover. The bird life had disappeared with the forest.

Nevertheless, its potential for restoration as a reserve for endangered native flora and fauna was enormous. In the first place its isolation from Bermuda's mainland meant that many of the introduced exotic species had not yet reached the island and could be prevented from doing so by quarantine measures. Notably absent from Nonsuch Island, when I first got to know it, were the entire range of naturalized tree species which are now dominant on Bermuda's mainland. Also absent were the introduced house mouse, toad and whistling frog. The anolis lizard had only just reached the island and was still rare. Although rats had also colonized by swimming from the main island I soon discovered that it was possible to eliminate them completely with an island-wide baiting program using Warfarin. One species which managed to survive on Nonsuch at precolonial levels of abundance was the endemic Bermuda rock lizard or skink., and a survey conducted on 1969 revealed that

it was twenty times more common on Nonsuch than on Bermuda's mainland.

The second feature of Nonsuch which made it so ideal as a reserve for endangered flora and fauna was its diverse topography, which made it possible to restore samples of all the major habitats of precolonial Bermuda except the wetlands. There was a surf beach with a dune; a sheltered cove beach; almost a mile of rugged coastline with cliffs and extensive tidepools; and a sheltered central vale with up to five feet depth of soil. There was even the potential for creating wetland habitats if minor modifications were made to the topography.

The cahow conservation programme was launched immediately after the discovery of the breeding islets in 1951. It soon became apparent why the cahow had escaped detection for so long. The bird is an extremely elusive and difficult subject to study. No one has ever photographed one on the open ocean, where the cahow spends most of its life. Returning from the ocean to breed only on the darkest and stormiest nights of winter, when its isolated and wave-swept nesting islets are least accessible to boats, it nests only in the deepest crevices of the rocky cliffs. It is impossible to see the nesting birds without the aid of a bright flashlight and a mirror attached to a pole. Cahows seldom leave any sign of their coming and going at nest entrances; even when they do, rain and salt spray soon obliterate the evidence.

It took ten years just to discover the entire nesting population of 18 pairs. By that time research had revealed that the major factor limiting the cahow was not rat predation, as previously suspected, but nest-site competition with the still common white-tailed tropicbird or longtail, whose breeding niche has always been the natural crevices and holes of Bermuda's coastal cliffs. In the absence of sufficient soil for digging their own burrows, two-thirds of the surviving cahows were trying to nest in coastal cliffs. This might not have mattered except that the breeding seasons of the two species overlapped. The winter-breeding cahows lay their single egg in January and are already leaving their newly hatched chick unattended in daytime when the larger and more aggressive tropicbirds return to breed in early March. These birds would simply peck the cahow chick to death, push it to one side and take over the nest for their own purposes. The adults of the two species rarely met, because by the time the tropicbirds laid and began staying overnight to incubate, the nocturnal cahows had abandoned their failed nest for the year. Both species are long-lived and faithful to the same nestsite, so the pattern would repeat itself year after year. It is incredible that the cahow could have survived for so long when two-thirds of the breeding pairs were being subjected to this loss every year.

The first major emphasis of the conservation programme, therefore, apart from rat control, was to solve this problem of nestsite competition. This was eventually achieved with a simple device called a baffler, which took advantage of the size difference between the two species. By fitting a board with a fixed-dimension, elliptical hole in the entrance to each crevice, the larger tropicbird could be excluded — just as starlings can be excluded from bluebird nestboxes. When this program was fully implemented by 1961, it effectively trebled the reproductive success and laid the foun-

**(Above)** The white-tailed tropicbird kills cahow chicks in competition for the cliffside burrows both species require. Manmade "Bafflers" make the cahows' burrows inaccessible to the larger birds.

**(Below)** Cahow condominums. On a rocky islet with no soil for the birds to do it themselves, a worker constructs "government housing." Cahows enter nests through the long tunnels. The circular holes in the roofs are for observation of each year's fledglings, and are covered with cement lids.

dation for recovery. No chicks have been lost to tropicbirds since that time.

I WAS AWAY FROM BERMUDA COMPLETING MY education during much of this time, but my first task when I returned from Cornell in 1957 to join the programme under Mowbray's direction, was to locate the entire population and make sure that all nestsites were fitted with bafflers. I then turned my attention to longer-term plans to accommodate the projected increase. Our research had indicated that the nestsite competition with the tropicbird was an artificially induced problem, caused by the displacement of the cahow from its optimum breeding niche. The accounts of the early settlers clearly indicated that on precolonial Bermuda, the cahow nested inland under the forest, in soil burrows excavated by the birds. It was the inland population that was most vulnerable to the introduced pigs, dogs, rats and cats and which was so rapidly destroyed. Only those few cahows which tried to nest on the peripheral offshore islands had any chance at all against the predators, but here they encountered increasing difficulty finding enough soil for burrowing, because of erosion due to the small size and exposure of the islets. More and more birds were forced to use natural crevices and holes, which brought them into direct conflict with the tropicbird.

In an attempt to recreate the original breeding niche separation I began constructing nesting burrows artificially on the vegetated tops of the nesting islets, where the tropicbirds were less likely to find them. This "government housing scheme", as I dubbed it, has gradually expanded until now more than half of the cahows depend on these artificial nest sites. They really had no choice because new pairs will colonize only close to pre-established pairs and there simply are no other suitable natural crevices on these islands.

While we realized that the population would continue to depend on man-made bafflers and artificial burrows for many decades, we were nevertheless anxious to lay the foundation for a time when the cahow could spread back onto one of the larger islands, with sufficient soil to enable the birds to dig their own burrows.

The most obvious island to accommodate such an expansion was Nonsuch because of its close proximity to the existing cahow islets and its isolation from the rest of Bermuda, making its management as a predator-free reserve feasible by quarantine. We finally persuaded the government to add Nonsuch to the sanctuary system by declaration in 1961. The following year one of the vacant and derelict quarantine hospital buildings was restored into a house, so I could move onto the island as a warden. I was newly married at the time and needed a place to live. Despite the hardship of island living — there was no electricity or telephone when I first moved out — I could hardly have dreamed of a greater paradise.

During that first year on the island, the concept of the Living Museum gradually evolved. I already knew that the cahow was unlikely to increase fast enough to spread back onto Nonsuch in my lifetime, because of its extremely low reproductive potential. Each pair produces only one egg a year, and only about half of these are fledged successfully. In addition, it takes eight to ten years before those fledglings reach breeding age, and during this time natural mortality reduces their number even further. By 1965 there were new and ominous signs that the breeding success was declining even further. The symptoms observed were identical to those being noted in populations of peregrines, ospreys, bald eagles and other predatory birds. After long and tedious international research the cause was eventually identified as the breakdown product of DDT poison in the environment.

The mode of DDT transport to the cahow was via the atmosphere and thence as fallout in rain to the ocean, downwind from the continent. Here, being fat soluble, it was immediately absorbed into the phytoplankton and thus into the foodchain where it became concentrated to dangerous levels by the process known as biological magnification. The cahow, feeding at the end of the foodchain, obtained just enough poison to cause eggshell thinning by enzymatic imbalances. The thin-shelled eggs were vulnerable to breakage in the nest, reducing breeding success from 60 percent to as low as 35 percent by 1967. The worst aspect of this problem was that it was international in scope and totally beyond my personal control.

The progress of the cahow breeding program was agonizingly slow; with the added menace of DDT, my efforts became a numbed routine. I needed a distraction, something with more hope of success, to justify Nonsuch Island's continued existence as a nature reserve. This is when it occurred to me to take advantage of its unique isolation and topography, and make it a sanctuary for all of Bermuda's terrestrial flora and fauna.

There was no money for restoration in the budget in those days. I was on a grant working on the cahows, and wasn't even employed by the government yet. The restoration of Nonsuch Island began as a spare-time diversion, but it quickly became my main project.

Work began on the Living Museum in 1963. The first task was to lay the foundation for restoring the native forest cover. In a sense I was starting with a clean slate, because when the cedars went the rest of the native forest and landbirds went with them. They formed the

**David Wingate with a boatload of native plant material headed for Nonsuch Island.**

(Top) 1975 — The restoration plantings slowly begin to resemble a forest. Faster-growing exotic species were used as perimeter windbreaks and later removed.

(Above) A Bermuda white-eyed vireo feeds nestlings in a casuarina shrub, planted as a windbreak.

backbone, the windbreak that protected all the other native species.

Artificially recreating the native forest required careful research, not only of the early historical descriptions and botanical surveys of the island, but also first-hand study of those few remaining, inaccessible corners of the island where remnants of the original flora had survived. Those remnants also supplied the seed of rare species for nursery propagation.

Between 1963 and 1972 I planted more than 8,000 trees and woody shrubs on Nonsuch Island, representing the full range of Bermuda's native forest species. It took a lot of persuasion to convince the government nursery to start raising its own native trees. Much of the plant material I had to propagate myself. Progress was painfully slow at first, because with no windbreaks severe winter gales and salt spray burned the little palmetto palms, olivewoods and other native trees right back to grass level. Even as late as 1969, Nonsuch barely looked any different, with the skeletons of the dead cedars still dominant above the grassy landscape. I began to fear that I would never even see the fruits of this labour in my lifetime. I was really doing something for the future, just hoping that the next generation would still believe in it.

In an effort to speed things up, I began using castor pomace fertilizer, and in 1967 I decided to plant the periphery of the island with a windbreak forest of casuarinas, the fast-growing, non-native, evergreen tree which was used extensively on mainland Bermuda for reforestation purposes after the loss of the cedars. While these two steps certainly accelerated the growth of my native forest, they also created some unexpected problems of their own. The tall-growing and wispy casuarinas could only be made effective as a windbreak by frequent topping, to make them dense and bushy like the cedar. This work was time-consuming and labour-intensive, and I began phasing the casuarinas out as soon as practical. The castor pomace fertilizer turned out to be an ideal high-protein food for the soil-burrowing, native landcrab, which began to multiply rapidly, causing soil erosion problems.

Nevertheless, the windbreak of casuarinas around the periphery of the island made it possible for the native forest to grow a little faster, and once it began to knit together into a thicket, the winter gales rode over the top. When that happened, the forest really began to take off, and I began to get some habitat diversity — microhabitats — and I could begin to seed in some of the more fragile understory species.

Throughout this period I had been unable to include the endemic Bermuda cedar in my reforestation, because the scale insects were still abundant and seedling trees were barely viable. However, approximately 4 percent of the mainland forest had managed to survive the scale, and by 1970 it had become apparent that natural selection, combined with effective biological control methods implemented by the government, were beginning to turn the tide. I finally undertook my first mass-planting of 600 Bermuda cedars on Nonsuch in 1972. Although death rates were high, approximately 200 ultimately survived and emerged above the other slow-growing native trees to dominate the canopy in several parts of the island.

BY THE EARLY SEVENTIES, I HAD AN EMERGING native forest, and it was now possible to begin thinking about restoring the native fauna which had lived in that forest. Some of the endemic components of that fauna had become extinct and were lost forever, but others, such as the local race of the white-eyed vireo, still persisted on Bermuda's mainland. In addition, a number of native species, though exterminated on Bermuda, were still potentially available for reintroduction from other parts of their range.

My first attempt at faunal introduction involved the Bermuda race of the white-eyed vireo. I had expected it to recolonize naturally from Bermuda's mainland, once the forest had become re-established on Nonsuch. Indeed, on several occasions I thought it had done so, when vireos were sighted on the island. By 1972 I had discovered that these birds were transients of the American race, which migrated between North and South America. The local race has greatly reduced flying ability, an evolutionary trend towards flightlessness common to many landbirds on mammal-free oceanic islands. It was so sedentary that it had failed to make the crossing over one mile of water. To overcome this problem I netted several of the vireos on the main island and released them on Nonsuch. They settled so successfully into the recreated environment that they now exhibit a population density twice that on Bermuda's mainland. They even seem to be reverting to their original fearlessness, as described by the early settlers.

My next attempt at a faunal reintroduction involved the green turtle, a species which still lived in Bermuda waters but which had been exterminated as a breeding species during the early years. In 1964 I wrote to Dr. Archie Carr, foremost authority on sea turtle conservation in the New World, to enquire whether he knew of any way to induce sea turtles to breed on Bermuda again. It turned out that he was conducting experiments on that very problem at the time, in a project known as "Operation Green Turtle". This involved the transplanting of batches of hatchlings from the last huge nesting colony at Tortuguero, Costa Rica, to beaches on former nesting islands where they had been exterminated. The experiment was based on the hypothesis that if the hatchlings enter the sea from a particular beach, they imprint on that beach and return there when mature. Between 1968 and 1978

a total of 16,000 green turtles were hatched and released from hatcheries on Nonsuch Island, South Beach dune and another nearby main island beach.

Unexpected developments before the end of the experiment cast doubt on the likelihood of success. Biologists in America discovered that the sex of green turtles is determined by the temperature of the sand during mid-incubation. A 2° centigrade variation makes the difference between all females and all males — males being produced at the lower temperature. We checked the temperatures of the hatchery beaches and were horrified to learn that they were generally well below those at which all eggs become males. This finding raises some interesting questions about Bermuda's precolonial sea turtle population and underscores the importance of preserving isolated genetic stocks or subspecies, on the basis that they may have evolved specific adaptations. It may well be that the original Bermuda population had the ability to achieve a balanced sex ratio at lower temperatures or was otherwise adapted to select sites where the correct ground temperature occurred.

In 1975 my dream of artificially creating wetland habitats on Nonsuch was finally realized with a grant from the New York Zoological Society. The U.S. Navy provided the loan of a landing barge for heavy equipment. No harm was caused to the developing native forest, because the sites for the two ponds had been reserved from the beginning and the pathways were just able to accommodate a bulldozer.

The small freshwater pond was created by slightly deepening a depression between two hills, laying down an impermeable plastic liner and covering it with soil. The liner formed a hanging water table, which trapped rainfall to produce a four-foot-deep pond. As soon as the average water level — determined by the equilibrium between rainfall and evaporation — was es-

**1975 — Constructing the artificial freshwater pond. A bulldozer scooped the pit and a plastic barrier was installed. Here it is covered with soil so aquatic plants can root.**

1977 — Two years after construction, the fresh-water lagoon is ready for use by wildlife. Unlike forests, aquatic restoration is rapid. A saltwater marsh was also constructed, completing the two native habitats of Bermuda that were missing from Nonsuch Island.

Landcrabs, scourge of tropical golf courses. When their natural enemy, the yellow-crowned night heron, was exterminated, the landcrabs became pests. Reintroduction of the birds provided biological control, eliminating the need for poisoning the crabs.

tablished, I planted the edge with the various native marshplants of Bermuda. Fish and invertebrates were introduced by transplanting buckets of water and mud. Within little over a year I had established a community that was indistinguishable from a natural marsh.

The beauty of wetland habitats for the restorationist is that you can create a mature community in just a couple of years. With the forest I'm still waiting — it's been 26 years and it is only a young mature forest.

The saltmarsh pond was created in a very low-lying area immediately behind the South Beach dune by simply excavating to below the water table level. In this area a pond resulted from natural seepage without the need for a liner. Once again the appropriate submergent plant and invertebrate communities were established by transplanting buckets of water and mud from main island salt marshes. In addition mangrove trees were planted around the perimeter and a rare endemic killifish was introduced from a small mainland pond.

With the completion of the two ponds, all of Bermuda's precolonial habitats were now represented on Nonsuch Island. Both of these ponds have since be-

come attractive habitats for a wide variety of migrant waterbirds, adding greatly to the interest and beauty of the island. A bird blind has been built at the edge of the freshwater pond to facilitate viewing without disturbance.

The ponds were a prerequisite for my next project in restoration, which was the reestablishment of a species that had been totally exterminated from Bermuda. The early settlers had described herons and egrets of several species, so tame that they could be clubbed down out of the trees. It wasn't long before they were completely exterminated. Although their nearest relations — migrant herons from North America — continue to be common as transients and winterers, they had never reestablished nesting colonies. For several years I had noticed that one of these species, the crustacean-eating yellow-crowned night heron, would eat my fertilizer-feeding landcrabs during stopovers on

Nonsuch. It occurred to me that if I could induce night herons to breed on Bermuda, they might serve as a valuable biological control for these crabs, which are generally regarded as a pest on lawns and golf courses.

The fossil discovery of a night heron skeleton in a Bermuda sinkhole at about this time confirmed the wisdom of my choice. The endemic Bermuda night heron had clearly been derived from the yellow-crown, but had evolved shorter legs and a heavier bill, adaptations specific to feeding on the heavily armoured terrestrial landcrabs.

To induce them to become residents, I decided to use the same technique that had been used with the green turtle, bringing in nestlings and weaning them into the wild on a diet of landcrabs. With Bermuda government funding and support as a landcrab biological control measure, I was able to obtain nestlings from a large rookery in Tampa Bay, Florida. From 1976 to 1978, a total of 44 yellow-crown nestlings were shipped to Nonsuch and reared in an abandoned building. Although hand-rearing of this species had never been attempted before, it proved to be the easiest and most successful of my restoration projects. Night herons feed their chicks by regurgitating into the nest rather than feeding each chick directly. This meant that I merely had to place the chopped-up crabs onto a food tray. Whenever the chicks were hungry they would gather around this tray like barnyard chickens. As soon as they were old enough to fly, I permitted them to escape from the building to learn to hunt on their own, but they continued to return to the food tray until they were proficient.

As I had hoped, these herons did not leave Bermuda, although they wandered extensively in the landcrab-infested areas beyond Nonsuch Island. As early as 1980 I confirmed successful nesting in the Walsingham Nature Reserve — a 40-acre wilderness tract on the opposite side of Castle Harbour. The population has continued to increase ever since.

At first I was disappointed that they did not nest on Nonsuch Island. They made good use of its ponds, often bringing their fledglings over to bathe and roost there as soon as they could fly. Then in 1985 I discovered that a small nesting colony had established in coastal buttonwood bushes on the isolated South Point of Nonsuch. As the native forest continued to grow, it gradually became more favourable for them. After a quarter-century, the palmettos, olivewoods and cedars were at last attaining maturity and beginning to self-seed. Everything was knitting together. My elation could hardly be contained when in 1987 I found night herons nesting throughout the island's

A yellow-crowned night heron nestling, transported from Florida.

(Below) Dinnertime for baby night herons; chopped landcrab was always on the menu.

forest, just as they must have done in precolonial time. They continued to feed almost entirely on landcrabs, reducing the Nonsuch population to manageable levels again and doing such an effective job on the golf courses on Bermuda's mainland that the managers were soon able to stop the use of poison baits for crab control.

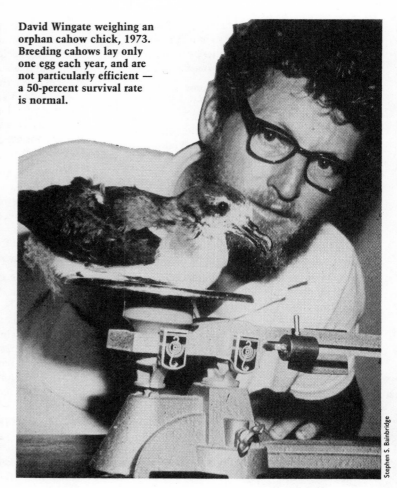

David Wingate weighing an orphan cahow chick, 1973. Breeding cahows lay only one egg each year, and are not particularly efficient — a 50-percent survival rate is normal.

Stephen S. Bainbridge

THE SUCCESS OF THE NIGHT heron reintroduction was the first to demonstrate that the Living Museum concept could have benefits for the rest of Bermuda, beyond the animals and plants living on Nonsuch Island. Others were soon to follow.

In 1982 a local marine archeologist and treasure diver, Teddy Tucker, returned from the Bahamas with a small collection of living topshells, *Cittarium pica*, an economically important food species of the intertidal zone throughout the West Indies. Early accounts record that this species was common in the waters of Bermuda too, providing an important source of food during the rat-induced famine of 1614 - 1618. This is confirmed by the abundance of empty shells, which can still be found on any shallow-water dive around the island. Overharvesting had probably tipped the balance to extinction, since this was the northern limit of the species' range. Eighty-two *Cittarium* were reintroduced into the intertidal zone of Nonsuch by arrangement with the Bermuda Fisheries Department in 1982. Survival and growth rates were excellent, despite occasional predation by octopus. By 1987 there were still 35 mature specimens and the first confirmation of successful reproduction was obtained. If this restocking experiment succeeds, a potentially important fisheries resource will have been restored as another economic benefit of the Living Museum. And the empty shells of *Cittarium* were once the home of the land hermitcrab, now rare on Bermuda. If the topshells become common again, these crabs may also multiply again and roam under the native forest canopy of Nonsuch, protected from the night herons by the heavy armour of their topshell homes.

As the native terrestrial ecosystem of Nonsuch Island has gradually begun to fit back together again like the pieces of a puzzle, the most important benefit has gradually become apparent — its value as an environmental education center for Bermuda's schoolchildren. Curiosity about the project has grown steadily and requests for field trips began as far back as the 1960s. As the project has matured, the island has become ever more diverse and interesting and its value as an aid in teaching about Bermuda's heritage has increased. By confining visitors to a system of grassy trailways and making use of blinds, it is possible to show all of Bermuda's terrestrial habitats in microcosm without disturbing the wildlife. Herons now nest within sight of the trail, giving a genuine feeling for the life of precolonial Bermuda.

The Living Museum continues to provide new insights, too. On September 25, 1987 Bermuda experienced its first hurricane in 25 years and the most severe winds and sea flooding in almost a century. In the 35 years since the loss of the native Bermuda cedars, an entirely new flora had developed on the main island through deliberate and natural reforestation. It was characterized by fast-growing and tall tree species, more than 95 percent of them non-native. This forest had never been tested by a major hurricane and the result, when it finally happened, was unmitigated disaster.

In less than three hours, Hurricane Emily destroyed more than 30 percent of the trees, mainly by uprooting and breakage. Ornamental parks and gardens dominated by the tall-growing casuarinas were especially hard hit, with up to 70 percent blow-down in some

A cahow chick. In 1988, twenty-two more were successfully fledged and set out to sea.

areas. Utility wires were devastated, leaving most of the island without electricity for two weeks. The vulnerability of this new non-native forest to wind damage had been predicted and forewarned against for several years, as local conservationists tried to encourage residents to make more use of native species in their gardens. But the warnings went unheeded until Emily. It only needed the benefit of a field trip to Nonsuch after that to put the point over forcefully, because its smaller-statured native forest survived virtually unscathed. The demand for native tree species for planting gardens and roadsides has since soared.

Indeed the only significant damage to the trees on Nonsuch Island occurred where I had earlier made some now-obvious mistakes in the habitat placement in my plantings. For example, I had overplanted with the coastal seagrapes, placing several of them too far inland from their natural coastal niche, but I had been reluctant to remove them artificially. Every one of these seagrape trees was blown down by Emily, giving me the opportunity to replant the correct species for that niche, which in this case was the cedar. Cedar was the pioneer tree of fresh blow-down or fire-damaged sites in precolonial Bermuda anyway, so I was merely emulating a natural process and edging my way towards a more genuine recreation of the past.

Emily began to provide some real character to the maturing forest by enhancing the wind-sheared and gnarled appearance of the trees, and by partly uprooting some palmettos, which will now record a pronounced bend in their trunks as they continue to grow. And there was a further lesson. I had often pondered how the cahows could have excavated their burrows into the root-matted soil under Bermuda's precolonial forest, because the soil on Nonsuch had remained hard-packed despite the shallow burrowing efforts of the landcrabs. But in a few moments of extreme violence, Emily had slightly uprooted a number of trees, lifting the root mats to create ready-made underground cavities ideal as nesting places.

This brings me back full circle to the cahow itself, that incredible symbol of survival and hope, which inspired the Living Museum project in the first place. Out on the offshore islets, the cahows' fortunes had begun to turn around. This resulted in part from the baffler-induced increase in productivity of the early 1960s which was beginning to be reflected eight to ten years later, as those extra chicks returned to breed. The battle to ban the use of DDT in the U.S. and Canada had finally been won and the breeding success of the affected species was gradually beginning to improve.

In the eighties the cahows experienced a further surge of population increase. By 1987, the population had doubled to 42 pairs. It was even resilient enough by then to absorb the loss of five sub-adult birds, killed by a vagrant snowy owl from the Arctic.

Throughout the sixties and seventies I would often walk down to the South Point on Nonsuch Island at the end of a hard day's work, hoping just once to be lucky enough to see or hear a cahow from there, without having to take a boat to the smaller islets. It finally happened one stormy night about 1982 and I have never been disappointed since. Indeed, I can now on occasion sit in comfort on the porch of the Nonsuch warden's residence and listen to the eerie calls of cahows out over the bay. I still don't know whether my dream of seeing them colonize under the island's restored native forest will be realized in my lifetime, but I might have good reason for cautious optimism — were it not for what is happening to the rest of my homeland across the other side of the harbour. ■

*In 1980, CoEvolution Quarterly (Whole Earth Review's predecessor) published an article entitled "The Subtlest of Catastrophes — Erosion vs. Reforestation at Auroville." It described a group of westerners who had started an intentional community in Tamil Nadu (southern India). They were battling severe erosion brought on by generations of subsistence agriculture. The restoration work described there continues, and what has emerged in addition to new forests is a successful model of First World/Third World partnership.*

*Alan Lithman is a twenty-year resident of Auroville. For information, write: Auroville International USA, P. O. Box 162489, Sacramento, CA 95816.*
*—Richard Nilsen*

# REVISITING AUROVILLE

## *A twenty-year-old project that's still growing*

### BY ALAN LITHMAN

**Monsoon rains on a denuded landscape produce severe gully erosion around Auroville.**

WE WERE dragging our bicycles across the barren fields, avoiding the sharp stubble, all that was left by the migrant herds of cows and goats. A merciless sun beat down upon this wretched piece of earth, bleaching it bone white or a brittle terra-cotta. A once-living earth dying back into a moon. We reached the edge of a canyon whose fingers gouged through the landscape. My friend pointed across the ravine to the barren plateau beyond, where a few palmyra trees shimmered like phantoms in the heat waves. "There it is," he said. "Auroville." I looked and saw nothing but a vacant landscape that slid into the Bay of Bengal. How could I possibly live there? How could anyone?

That was in 1969. Today, two million trees later, standing at that same rim of the canyon, forests camouflage the horizon and the habitations, schools and workplaces where some 700 men, women and children from more than 25 nations are learning to live a sustainable future together.

Transplanting a growing microcosm of Westerners, plus Indians from diverse socio-economic backgrounds, onto a plateau in rural southern India (already inhabited by some 20,000 Tamils in a dozen villages) was a unique experiment. In the past, such a venture was called colonialism. How was Auroville, with its ideals and dreams of the late sixties, to take root in such a reality without (well-intentioned or otherwise) taking it over?

That is a question we are still dealing with. When you look at the bare earth, it does not lie. There is simply nothing more to take. If anything at all is to grow on this plateau, we must put everything back: the soil, the water, the life. Everything — the evolutionary capital that a century of human ignorance and greed squandered. There comes a moment when matter itself, despite our machinations and abstractions, our rhetoric and polemics, makes its point. When the earth speaks. And in this sense, we have all been colonialists on this planet. At Auroville, there was simply no buffer with which to fool ourselves. It was clear what had to be done, and there was no one else to do it for us. We were humanity coming home to repay a terrestrial debt from the West to the East, a karma that we owed to the Earth.

**(Above left) A view of the plateau from a farmhouse window in 1976. (Above right) The same view in 1986 shows a well-established forest of indigenous trees.**

So we began. A growing core of Aurovillians — "green-workers," we later called ourselves — fanning out across a landscape, challenging its stark emptiness with crowbars and mattocks. An absurd image, laughable even to the Tamil villager who could not fathom the sense or ulterior motive for such a ritual in this utter wasteland. With our crowbars, we broke through the exposed clay crust whose topsoil had long ago washed away. And with our crude handshovels, *mumptis*, we dug pits in the dense red clay. Into this sterile ground whose surface we could barely scratch, we began to return what had been taken away. We carted compost and topsoil to fill the pits, and hauled barrels of water by bullock cart to the seedlings we planted.

Who could imagine in those first impossible days what would grow in the years and decades to come? We certainly knew no more what we were about than the villager who eyed us curiously from his field, as if we had just landed from another planet. But we persisted, quick to learn from our discoveries and failures, encouraged by nature's response to those first efforts, inspired by her resilience and the amazing speed of the regenerative capacities of the tropics.

As we persisted, certain things became obvious, such as initiating the reforestation with hardy varieties of indigenous, drought-resistant trees and scrub jungle to temper the environment before introducing slower-growing stands that would have withered in the open sun; providing trees with natural fencing or watchmen to protect them from the foragings of cows and goats or the sharp *kutti* of the villager in search of firewood; planting just before the monsoon, to allow the seedlings a two-month soaking to establish themselves; and raising the earth in a circular mound around each pit, then mulching it to retain the moisture.

Quick-fix aid programs generally overlook these essentials and the follow-up care necessary to assure significant, healthy survival. It is painful to see thousands of seedlings stuck in a field, watered once and abandoned. The contractor has fulfilled his obligation, signed the paper satisfying the donor agency that so many thousand trees were planted. The fact that none survived is not mentioned at the Board Meeting.

To support the continuing growth of Auroville's trees, an

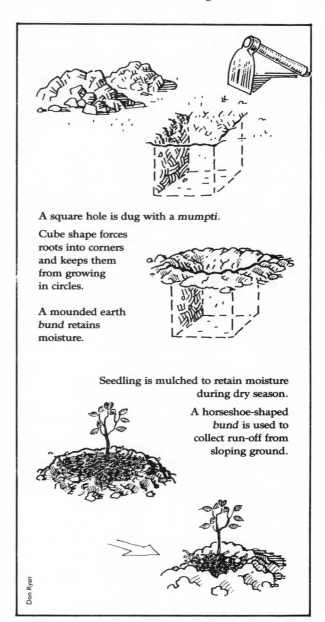

A square hole is dug with a *mumpti*.

**Cube shape forces roots into corners and keeps them from growing in circles.**

**A mounded earth *bund* retains moisture.**

**Seedling is mulched to retain moisture during dry season.**

**A horseshoe-shaped *bund* is used to collect run-off from sloping ground.**

Don Ryan

(**Right**) Tall trees provide shade for the nursery, which propagates indigenous varieties plus endangered species from throughout India and Southeast Asia. (**Below**) Tending plants on a raised earth bund. Cactus is used as natural fencing to keep livestock away from young trees.

entire infrastructure emerged. Nurseries sprang up to provide the seedlings and recultivate endangered species. Large-scale composting areas were developed to revitalize the depleted soil. Extensive networks of check dams in canyons and miles of earthmound grids (*bunds*) around fields were devised to retain the deluge of monsoon rain that would otherwise sheet-flood across the plateau, sweeping off tons of precious topsoil with it to the sea. Water systems, powered primarily by scores of windmills designed and built in Auroville, were situated at strategic locations throughout the young plantations to provide gravity-feed pumping stations for the bullock carts that still must haul water to the trees. Greenwork meetings became a regular planning exercise in which fieldworkers could share experience, research and strategies, as well as reach out beyond Auroville to other regions of India.

It became more and more apparent that Auroville, with its present patchwork of 2,200 acres spread out over a larger area of 12,000 acres, could not realize its true potential in isolation, either environmentally or on any other level. After all, what was the purpose of Auroville? To become an exclusive patch of green in a parched corner of India on a planet whose forests are being felled at the rate of a hundred acres a minute? To be an eco-elite, adding twenty minutes worth of earthtrees while the villager next door cuts the sole remaining acacia in his field for firewood? Each season that same villager, whitened with the dust of the DDT he threw by hand into his overworked field, would squeeze a little less out of that plot of earth, and the water table of the entire region would continue to fall.

For Auroville, an effective change required a comprehensive and cooperative action plan for the plateau as a whole. To check the monsoon's fury as it plunders the slope, turning the rutted tracks of bullock carts into ravines, *all* of the fields must be protected. Self-interest no longer stops at the boundary marker, as we are beginning to learn in global terms as well.

Yet with villagers for whom the land literally means survival, abstract environmental arguments do not feed people. Turning around the momentum of such self-destructive patterns requires a different form of communication: action, demonstration, patience and time.

After two decades and two million trees, something does translate; simple things have become tangible. There is a change in the microclimate, the soil fertility, the watershed. Perhaps a well in the village that had gone dry suddenly has water again. Why?

Slowly, villagers are beginning to assimilate the impact of Auroville's action and to draw their own conclusions. Raised earth grids are appearing around their fields along with checkdams in the canyons that threaten them. Trees are replacing some field crops. In response, Auroville provides whatever support and assistance it can — training programs, techniques, seedlings, even joint proposals for funds to restore the balance of this piece of earth we share.

The circle continues to widen as Auroville initiates outreach programs in other parts of India, pressing beyond state borders toward a true First World-Third World partnership.

Looking back across the threshold of that canyon I crossed twenty years ago, I see a handful of naive humanity that dared to begin this adventure with crowbars and handshovels. And I see what is possible when we stand our ground, our common ground. I see forests and grasslands filled with masses of flowers and the native birds and wildlife that had long ago disappeared from this part of the planet. I see what can be done from the barest beginnings and under the most impossible conditions, with hardly any means or resources. Not by calculating, or waiting for the opportune moment, or the big money, or for a conference to confirm what must be done. I see what can be done by the power of simply doing it. And as I turn toward the starkly contrasting landscape behind me, I see all that is yet to be done. ■

# Australia Turns An Environmental Corner

## BY RICHARD NILSEN

Australia made history in 1989 when Prime Minister Hawke faced up to his nation's environmental deterioration and, in a statement titled *Our Country, Our Future,* called for a national policy that includes restoration. It's the kind of high-profile pronouncement that many people in the U.S. have long wanted. Consider it a bold admission and a promising first step, while huge issues remain undecided, like the fate of Australian forests, and whether to allow mining inside national parks. It's hard to know from this distance what is real and what is rhetoric, but the problems Australia faces are not in dispute.

The continent's isolation made what has happened all the more severe. The Aborigines brought the dingo with them into this marsupial paradise about 30,000 years ago, but that was as nothing compared to the European settlement just 200 years ago. Sheep and cattle, yes, but also rats, cats, rabbits and foxes. Not to mention escaped exotic plants since then,

Victorian Salinity Program

like a Central American legume called prickly mimosa that's overrunning the bush of North Australia, or South African capeweed, which has infested areas equivalent in size to Arizona *and* New Mexico.

One of the prime minister's proposals is to plant one billion trees by the year 2000, of which many will be going into the Murray-Darling Basin, where problems with soil salts have become critical. This drainage has half of Australia's sheep and croplands, and provides half of South Australia's water. Farmers cut the native forests

Water samples are collected by students during Saltwatch week, from a groundwater bore in the Mallee in the northwest of Victoria.

to plant crops, and this upset a delicate hydrological balance. The salt-bearing soils, underlain by an impermeable layer of clay, waterlogged, and the soluble salts rose to the surface, where they are ruining farmland and poisoning rivers. Lowering the water table is the name of the game now, hence the tree-planting.

Some of the tactics for raising public consciousness are worth noting by residents of the western U.S. facing similar problems. The official Salinity Program in Victoria dumped seven tons of salt — the amount entering the Murray River every three minutes — in a Melbourne city square. There's a Saltwatch program with stickers that say "Involve Me and I'll Understand"; there's a newsletter called *Salt Force News.* There are also thousands of citizens collecting water samples for a database, some from borehole wells that keep tabs on the water table in public places like shopping-center parking lots and schools. From these wells emerge color-coded rods attached to floats. If the water table rises, the rod shows red, and everyone gets the message. ■

## Case Studies in Environmental Hope

*This "accentuate the positive" collection of eighteen case studies in environmental progress from Western Australia is the kind of effort that every bioregion on the planet should be publishing. Buy one and use it as a model.*
—Richard Nilsen
[Suggested by Stephen Hodgkin]

### Case Studies in Environmental Hope
Peter Newman, Simon Neville, and Louise Duxbury, Editors

**$13.50** postpaid from Australian Book Source, 1309 Redwood Lane, Davis, CA 95616; 916/753-1519

●

*Rolly Carrol of Westralian Sands, and his tree tubes.*

Following sandmining, the deep sands are

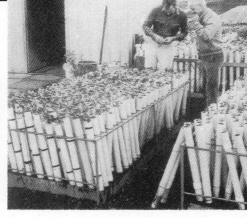

left in an almost unconsolidated form, and are very difficult to rehabilitate. Rolly reasoned that if a plant had very deep roots and very few leaves, it would stand a greater chance of survival than normal seedling stock. He developed a method

using split PVC tubing, and now produces plants which can be planted all year round, even in paspalum or kikuyu grasses, and have a very good chance of out-competing the weeds and flourishing.

# NO LONGER FLOWS THE NEVA

## Leningrad Grapples With A Crisis In Water Quality

### BY PETER RIGGS

Sergei Tsvetkov

**A**NDREI SHIFTED INTO low gear
as the heavy truck climbed the steep
grade to the causeway. "They'll never
stop building this dam, you know,"
he said, shouting over the roar of
the diesel engine. "They'll keep building it right up
to the day they begin tearing it down." The truck
reached the gravel road at the top of the dam. To the
left, the cranes and shipyards of Leningrad's harbor
were plainly visible under a slate-grey sky. Beyond
the dam's perimeter lay only the Bay of Finland, the
easternmost finger of the Baltic Sea. "All this time
building," Andrei continued, "and still nobody
knows for sure whether the good in this project will
outweigh the bad. But I'll tell you this — every
summer, when my boys and I go swimming, we have
to go farther and farther away from this causeway,
because the beaches near it aren't clean anymore."

At daybreak here in the old Czarist capital, sleepy-eyed
women in housecoats and felt boots haul tin buckets to
curbside and join a long queue, anxiously awaiting delivery of yet another scarce commodity — drinking water.
Fresh water must be trucked into the city, as the oily-
brown liquid which flows through Leningrad's taps is
"categorically not recommended for human consump-
tion."

The beauty of Leningrad — proudly called "Peter" by the
city's five million residents — is also its curse. Czar Peter
the Great constructed his capital far from Moscow so that
it could serve as a "window on Europe," and even today,
Leningrad retains a strong European flavor. But the delicate inlay of canals and watercourses also testifies to the
fact that Leningrad is built on an estuary — an environment extraordinarily susceptible to ecological disruption.

Water pollution in Leningrad, in the words of one activist,
"is at catastrophic levels; people have no idea how close
we are to a breakdown of the entire ecosystem." The Neva
River and its tributaries have been badly damaged by industrial wastes, agricultural runoff, and the installation

---

*Soviet environmental problems are just beginning to emerge into public view on both sides of what used to be called the Iron Curtain.
The Eastern Bloc is (appropriately) preoccupied with political restoration. Will the environmental variety be far behind? The plot of this
Russian tale is familiar the world over: citizen activists versus the bungling bureaucrats and their big dam. One good effect of the environmental crisis is that it melts national and ideological boundaries. While politicians on both sides scramble to keep up, history is
being made by people.*

*Peter Riggs is an American teacher and writer currently living in Indonesia.*

*—Richard Nilsen*

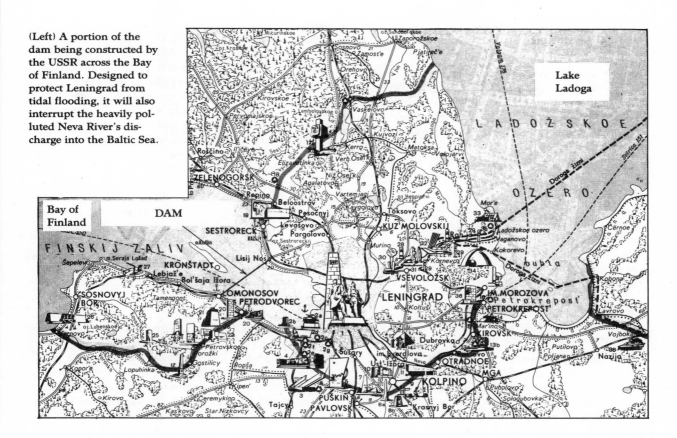

(Left) A portion of the dam being constructed by the USSR across the Bay of Finland. Designed to protect Leningrad from tidal flooding, it will also interrupt the heavily polluted Neva River's discharge into the Baltic Sea.

of new hydrostations which strip the waters of their self-purifying capabilities; effluents from the numerous paper mills and chemical plants at nearby Lake Ladoga have resulted in outbreaks of disease from water-borne bacteria; and the shallow marine ecosystem of the Bay of Finland, unable to assimilate Leningrad's untreated wastes, is being rapidly degraded.

A dam now being constructed in the delta to protect Leningrad from tidal flooding will greatly exacerbate the problems. These "protective structures" would sever the link between the Neva River and the Bay of Finland, transforming the mouth of the Neva into "a cesspool of unpurified sewage," according to the activist.

The "protective structures" will stretch twenty-five kilometers across the bay, between the beachfront suburbs of Lisij Nos ("Foxnose") and Lomonosov. The dam is now a little more than half-finished; construction costs were underestimated, and the timetable for its completion was completely unrealistic. Although in 1976 the all-union Council of Ministers approved a set of measures calling for the installation of treatment facilities at 100 percent of Leningrad's wastewater outlets by 1985, and noted at that time that it would be necessary to implement this clean-up before the dam could be completed, the number of untreated wastewater outlets actually *increased* in the decade up to the 1985 deadline, and funds earmarked for the smaller-scale water treatment programs were diverted to the dam.

The structures are a colossal monument to "gigantoma-

nia" — a disease not unique to but certainly prevalent in the Soviet Union's production ministries, powerful bureaucracies well-accustomed to writing their own meal tickets. The dam itself is built to withstand a flood likely to occur only once in ten thousand years. Scientists have calculated that a meteorite crashing on Leningrad's famous Nevsky Prospekt is equally likely.

The dam is not the first scheme designed to protect Leningrad from floods, however. In the early post-war period, plans were drawn up to sink an atomic bomb in the bay near Sosnovij Bor ("Pine Woods"); in case of flood, went the reasoning, the bomb could be detonated and the extra water would flood into the crater resulting from the explosion. When asked about the resulting fallout, a ministry spokesman was said to have responded, "What do you think is better — to drown, or to brave the slow effects of radiation, which will gradually disappear?" One Leningrad activist jokes that, instead, residents of "Peter" have the worst of both worlds: "Because of the dam, we will drown slowly in our own shit, which is gradually accumulating."

**T**HE SOVIET UNION, which for many years insisted that pollution was a phenomenon found only in capitalist societies, has finally acknowledged the extent and scope of its environmental problems. Bureaucracy, the overriding importance of production-plan fulfillment, and Soviet citizens' ingrained pessimism towards the possibility of change has thus far hampered the formation of successful responses to such problems.

Now, however, the prevailing climate of glasnost coupled with the growing political sophistication of grassroots organizers throughout the Soviet Union has given rise to a powerful Soviet environmental movement with both a local and a global focus. Because of its own precarious ecological situation and its well-established contacts with events and movements around the world, Leningrad has the best-developed environmental movement of any city in the Soviet Union.

The Soviet Union's ecology movement has come of age at a time of increased awareness of the global nature of environmental problems and easier transborder communication. Until glasnost, environmental problems in the Soviet Union were seen as the concern of specialists only; now unofficial environmental groups in cities like Leningrad have ended the state monopoly on environmental information by learning to use computer links and databases to keep abreast of scientific and political developments on the local and international level.

Other tactics employed by independent environmental groups in Leningrad will also be familiar to western activists. Last year the group Epicentre organized a three-day "teach-in" about Leningrad's water system; members of the group Adelaida have conducted a continuous signature campaign calling for a halt to dam construction; and members of the Ecological Committee of the Leningrad Writers Union have compiled a number of scientific studies which warn of the grave threat to public health posed by unchecked pollution of the Ladoga-Neva drainage basin.

Sergei Tsvetkov, chairman of the Ecological Committee, remarks that "We now have a vigorous group of experts helping us sift through the great quantities of data pertaining to the dam project, as well as epidemiological studies from Lake Ladoga. There are several ministries with a powerful interest in seeing the dam built; they employ scientists to make studies showing that the negative environmental impact will be minimal. We too have enlisted the support of about one hundred scientists from various fields; in this way we can use facts to counter the statements from the production-oriented ministries, which is crucial."

Links with scientists and grassroots groups abroad have already yielded some tangible results for Leningrad activists. As scientists Boris Gusakov and Nina Petrova began to wade through epidemiological studies from Lake Ladoga in an effort to understand the anthropomorphic changes taking place in this lake, data sets from abroad saved them from unnecessarily duplicating numerous hydrological and biological studies. Much of this research had been carried out at Lake Erie — from the first "heyday" of American environmentalism, when ecology became a critical issue because of the degradation of the Great Lakes.

Eventually Gusakov and Petrova found the parallels so striking that they decided to publish a book. *The Great Lakes Face to Face* compares the Erie-Ontario system with that of Ladoga and Onega, the two large lakes whose waters flow through Leningrad into the Bay of Finland. There are literally hundreds of polluting industries located at Ladoga; the Syakij paper-cellulose plant alone discharges more than seventy million cubic meters of wastewater into the lake. The book explores the similarities of the ecosystems, the patterns of human use of the lakes' resources, and the clean-up projects initiated in both the United States and the Soviet Union. The authors conclude with a call for greater interawareness: "Information obtained by scientists must be translated; information must give shape to practical recommendations and social-economic valuation. All people from both continents must safeguard the unique resources of the North American Great Lakes — and of our own, great lakes."

UNTIL RECENTLY, chairman Yuri Sevenard and other members of the interministerial committee overseeing construction of the dam were able to suppress publication of information which questioned their findings. Now that environmental activists have access to a wider variety of data and are able to publish relevant articles in the official press, they are less easily cowed. But no less than anywhere else in the Soviet Union, glasnost in Leningrad is proceeding by fits and starts.

"We feel that the TV and radio people are just playing with us; they are not willing to engage in a serious discussion," says Sergei Tsvetkov. "The city and regional authorities use the press to give the impression that there are no conflicts, that because we've got glasnost everything will work out just fine." Still, most bureaucrats are finding that the glare of publicity is not always to their liking; in many cases, they have resorted to more familiar strong-arm tactics in order to get their way.

For example, producers of the Leningrad television program "Public Opinion" planned to broadcast live a "citizens' discussion" which they hoped would "create a milieu hospitable to the activization of new attitudes." Instead, angry crowds, demanding a complete accounting of the environmental effects of the dam project and the pollution of Lake Ladoga, converged on the site of the discussion; during the program attention was repeatedly diverted away from sensitive issues. Noted *Mercury*, the newsletter of Leningrad's ecology movement: "Television failed to keep its promises."

TASS felt obligated to respond. Their comment: "As is always the case surrounding tough questions, the social mood underwent some change. . . . In previous years only 'wholehearted, unanimous support' was noted, when what was really needed was the firm courage of citizens who would dare to come forward and speak against the project. Now it is easier to confess one's opposition to the project, and it takes a brave man to speak in support of this important undertaking." Environmental writer Oleg Petrichenko replied bitterly, "If formerly

Sergei Tsvetkov

Sergei Tsvetkov taking water samples from the Bay of Finland, within the confines of the dam. Water quality and temperature data are essential for calculating impacts like algal bloom vectors. This information is being used to counter the claims made by the dam-building ministries. Both East and West, those who control the data often will influence the final outcome.

# Perestroika And The Rule Of Law

One pervasive aspect of *perestroika* that has largely escaped Western attention — in part because it is only manifested through other issues — is the idea of the "Rule of Law." As the Soviet Union moves slowly towards a "civil" society — that is, one with a codified system of rights and responsibilities, where decisions reflect protocol rather than just the whims of the powerful personalities involved — the idea of accountability under the rule of law has been extended beyond just budgetary matters. Political scores are no longer to be settled in extra-institutional ways; those involved in the Tbilisi massacre, for example, are said to be "brought to account under the rule of law." In the environmental realm, firms and ministries for the first time must be prepared to thoroughly justify not only their use of particular natural resources but also the productive processes employed. No longer is production in and of itself hailed as an unassailable good.

This is of great significance for environmental activists in the Soviet Union. One of the most aggravating difficulties they have faced is that *on paper* the Soviet Union has one of the most stringent and comprehensive systems of pollution regulation of any country in the world. The existence of this legal framework for the protection of nature thus served to defuse effective criticism of policy-makers. But there were no provisions for enforcement. The legal framework was viewed by plant managers as a set of guidelines with which to voluntarily comply — up to the point where some provision or another hampered production norms. Fines levied against a firm for repeated transgressions were purely token — rarely more than 300 roubles.

—*Peter Riggs*

'critics' were threatened with the straitjacket for criminal insanity, now they're given the dunce cap as seekers of cheap popularity. The dunce cap is a little easier to wear, but it gives you about the same prestige."

A "public meeting" held on January 23, 1989, in the A. M. Gorky House of Culture, provides another example of bureaucratic railroading. On the agenda was the proposal for joint-venture construction of a Russian-style amusement park at the Lisij Nos end of the tidal dam. Experts gave their opinions, and bureaucrats gushed about the unalloyed good that such a "gigantic entertainment complex" would bring to the people of Leningrad. When, however, opponents of the plan rose to speak, the moderator switched off the public-address system and declared the meeting adjourned. This led to a signature-writing campaign, and the opposition group was able to

publish its letter in the progressive journal *Ogonyok*, to the great embarrassment of those bureaucrats involved. Now it is not clear whether construction of the amusement park will be approved or not.

In the face of such bureaucratic inertia, Leningrad's independent environmental groups have proven to be superb grassroots organizers, able to turn out support at critical times on a number of key issues. They have consistently outflanked state efforts to co-opt their activities. While Communist Party officials have constantly lamented the unwillingness of independent groups to link their efforts with those of official organizations, grassroots organizers point out that "people have a bad attitude towards state-sponsored groups. As far as we're concerned, the official Nature Protection Organization is just for show. It is under the thumb of the production ministries; it has given approval to all kinds of environmentally destructive ventures, including the dam project." The Party has been put in the paradoxical position of having to relearn grassroots organizing techniques in order to drum up support for the Nature Protection Organization. Most feel the effort to be too little, too late.

The Soviet Council of Ministers' January 1988 resolution on the environment, which called for the establishment of a State Committee for Environmental Protection, has been greeted with similar skepticism. "It's a Moscow-based initiative, and so far the efforts of this committee have been little more than window-dressing. We face a unique set of circumstances here, and only we ourselves will be able to effect change," commented a member of Delta, the grassroots organization fighting construction of the dam.

Delta took the initiative last spring in organizing a conference on the health of the Baltic Sea, which was attended by environmentalists from the three Baltic Republics, as well as representatives from Finland and Sweden. Closed beaches in Estonia, Latvia, and Lithuania in the summer of 1989 contributed to the feeling of malaise towards Moscow's rule, and "green" ecological programs are an important part of the independence movements in these republics.

GORBACHEV'S REFORM program has brought forth bold new initiatives and bitter backlashes. Behind the various calls for greater autonomy, for political freedoms and economic restructuring, is a question so basic that it is rarely verbalized: how many long-buried but still painful issues must be dredged up from the first seventy years of the Soviet Union's history before we can truly move forward? To what extent must past wrongs be redressed?

This is a question of crucial importance in the field of ecology as well, because it asks whether projects approved in Brezhnev's time — the "period of stagnation" — should be cancelled, at the loss of hundreds of millions of roubles in sunk costs. It is one thing to forbid construction of new nuclear power plants in the Ukraine, or new pulp mills at Ladoga; it is quite another to close existing facilities.

In this sense, the people of Leningrad face far more daunting tasks than those fighting only further development. While there is an element of "Not-In-My-Back-Yard" thinking in the program of Leningrad environmentalists — the debate over construction of the amusement park is one example — in general, Leningrad environmentalists are looking towards environmental restoration.

There is a creeping awareness among the people of this city that they are on the brink of disaster. Even with a complete ban on polluting activities at the lake, the Ladoga ecosystem will need at least thirty years to heal itself. Thus people feel that they must begin to redress environmental wrongs at the lake now. They argue that the "protective structures" severing the Neva River from the

---

# Soviet-American Environmental Exchange

We are hearing a lot these days about the triumph of capitalism and the failure of socialism: While that may be true in an economic sense, it is irrelevant in environmental terms — *both* systems are failing. The US and USSR have needed each other to keep their Cold War economies going all these years, and they need each other now, to begin cleaning up global pollution. While Soviets clamor for consumer goods and American defense industries begin to calculate the economic impact of sizeable cutbacks, this coalescing of environmental work and economic redirection has yet to emerge. The heady changes coming down in the Eastern Bloc make it all the more frustrating to watch the facile posing by the insipid airheads currently in Washington.

Meanwhile, realizing perhaps that it is better to be the thin edge of the wedge than to worry about airheads, the Soviet-American Environmental Exchange has formed. Taking full advantage of the new political climate, it has grown out of a trip to Leningrad, Vilnius and Moscow by 25 American environmentalists in March 1989. Its aim is to facilitate cooperation between both countries' non-governmental environmental organizations. The US and the USSR together use a majority of the world's resources, and are also the two largest polluters. On the long road toward the restoration of the global environment, this group might turn out to be an important player.

—Richard Nilsen

*Information free from: Center for US-USSR Initiatives, 3268 Sacramento Street, San Francisco, CA 94115.*

---

sea should be dismantled immediately, so that on a summer day families can once again swim in the shallow Bay of Finland. As Oleg Petrichenko remarks, "The Leningrad Dam is a fitting monument to our period of stagnation, built according to the rules of that time: without paying heed to public opinion; with money taken from the people. The millions already spent won't return . . . but any further funds should be used for the defense of Ladoga and the Neva River."

In any event, the various groups that make up Leningrad's "environmental opposition" will continue to press their case through the forums provided by the slowly expanding notion of glasnost. "We are not just fighting for public health, or the protection of nature," concluded Sergei Tsvetkov. "We are also working to change the nature of the political dialogue about environmental protection here, to legitimize the idea that a frank discussion of the environmental impacts of our lifestyle is urgently needed." ∎

Kit Miller

Brazil nuts are one rain-forest product that offers an alternative to burning trees to produce crops or cattle. Here they are loaded onto a barge bound downriver to Belem and shipment overseas. Rio Branco, Acre, Brazil.

# Letting The Amazon Pay Its Own Way

*BY JON CHRISTENSEN*

*It's a bit difficult to criticize forestry practices in other countries when we Americans are allowing the last sizeable temperate-zone rain forest on the planet (the Tongass National Forest in Alaska's Panhandle) to be turned into wood pulp to make Japanese rayon. There, as in tropical rain forests around the world, the future turns more on people and jobs than it does on trees and biomes. Solutions that are sustainable both for people and forests have received little attention so far, but without them, no preservation efforts will be successful in the long run.*

*Jon Christensen is a correspondent for Pacific News Service. He spent 1989 on assignment in Brazil with photographer Kit Miller, where they drove 18,000 miles and crossed the Amazon Basin from east to west and north to south.* —Richard Nilsen

I spent last year in Brazil, the better part of it in the Amazon Basin. I recently returned home to Nevada and the Great Basin of the United States.

I was concerned by what I saw in the rain forest, but somehow the High Desert Rain Forest Action Group, which I saw listed in a roundup of this year's environmental causes in the local paper, strikes me as too much of an oxymoron. At some point "think globally, act locally" starts to sound a little absurd.

One local rain-forest outfit is trying to get people to adopt trees in the Amazon. The director promises to send me a picture of my tree, somewhere in the Amazon, with a plaque inscribed with my name on it. He says it's a good way to guarantee my supply of oxygen.

In the box of junk mail awaiting my return, I found a whole slew of organizations touting ways I could help "save the rain forest" by writing them a check. I read about debt-for-nature swaps, ecological preserves and international bank campaigns.

Then I read in a back issue of *Nature* magazine that scientists have already come up with a solution to the burning problem of the year. They studied two and a half acres of Amazon forest and found out that harvesting fruits, nuts, gums, and oils is more profitable than logging or cattle ranching. The Amazon can pay its own way.

What a relief, I thought, as I burned the alerts and appeals in my fireplace for heat and read on.

Charles Peters of the New York Botanical Gardens paddled upriver with another botanist and a forester to inventory

Rubber tappers watch as an itinerant merchant weighs the rubber they are trading him for supplies from town. Jurua River, Acre, Brazil.

a plot of jungle 20 miles from Iquitos, Peru. In the small village of Mishana they talked to the locals to find out what would sell in the local market.

On their small plot, about the size of a suburban lot, they found 842 trees belonging to 275 different species. Despite this riot of life forms, less than half of the trees produced a marketable product. Still, Dr. Peters calculated that leaving the forest standing was a better investment than cutting it down for wood or clearing it for cows. Over one year, he calculated, your average savvy forest dweller could make $697.79 by harvesting nine varieties of fruit, wild chocolate, rubber, and the occasional tree from that little bit of jungle.

In one clear cut of the timber, Dr. Peters figured he might gross $1,000. But in the long run, that wouldn't be a good financial strategy. Calculating the cumulative value of sustainable harvests over 50 years, the researchers discovered that their piece of land was worth $6,280. A nearby tree plantation of the same size was valued at $3,184 while cattle pasture fetched only $2,960.

(These figures were calculated as a current value much as a financial manager would determine the value of a long-term bond. Although how researchers were going to get your average forest dweller to buy that proposition, when he doesn't know how to read or make simple calculations, was never explained.)

"Deforestation is a bad investment," Dr. Peters concluded. The most immediate and profitable way of combining the often-contradictory goals of development and conservation, he said, is the exploitation of the natural fruits of the forest.

He rejected the notion of setting up nature preserves in the Amazon. "I don't see any future in simply setting land aside as pristine parks," he told the *Washington Post*. "Not in the Third World. The local people won't buy it and the local governments won't buy it. The forests must be used in order to be saved."

The finding was hailed as "the last great hope for saving the rapidly disappearing rain forests." Environmentalists have long contended that the forest was more valuable standing than down, but now scientists had proved it in dollars and cents. Even the stolid *Economist* jumped on the bandwagon, enthusiastically endorsing the idea that free-market forces should decide the fate of the forests.

Then I noticed that products seeking to tap a rich vein of rain-forest concern and growing green consumerism are already appearing on store shelves. This year Ben & Jerry's launched "Rain Forest Crunch" with Brazil nuts gathered by rubber tappers in the Amazon. One can now "eat ice cream and save the rain forest."

"The idea is to stop saying no to everything in the Amazon and start offering a positive alternative to the people who need to make a living there," said Jason Clay, an anthropologist-turned-entrepreneur with the indigenous-rights group Cultural Survival. Early in 1989, Clay primed the rain-forest marketing trend by sending 880 pounds of samples from the Amazon to four companies in the United States and England, including 45 different fruits, nuts, oils, and flours.

The Body Shop, a booming chain of natural health and beauty stores, will soon be offering a Brazil-nut facial scrub. There is even talk of marketing natural-latex condoms with the slogan "protect yourself, protect the rain forest."

These green businesses want to help improve the terms of the Amazon economy for the poorest, most powerless forest dwellers — rubber tappers and Indians. By short-circuiting the local elite, which now controls most trade in forest products, they hope to return a bigger share of profits to ecologically sound projects in the Amazon aimed at diversifying and increasing production, and doing some processing on-site to capture more of the value locally.

When I was in the Amazon, the rubber tappers were promoting the concept of "extractive reserves" to protect the forest from speculation and deforestation. Chico Mendes died in the fight to establish a reserve in the state of Acre.

"Now we're talking — think globally, act locally," I said to myself as I hopped into my gas-guzzler and drove to the supermarket to scoop up a quart of ice cream. Back at home, my boots in the fire and a spoon in the Ben & Jerry's, I learned of another interesting plan to save the rain forests — by raising iguanas rather than cows. I almost spilled "Rain Forest Crunch" all over myself.

"Lizards," I thought, "Ick."

But I knew that cows are the main culprits behind deforestation; they occupy 85 percent of the cleared land in Amazonia. Then I remembered the crocodile we ate with a family of rubber tappers on a journey up a remote river. It turns out iguanas are also considered a delicacy in parts of Latin America. They call them "gallina de palo," or chicken on a stick.

The Green Iguana Foundation and the Smithsonian Tropical Research Institute in Costa Rica extoll the efficiency with which the sedentary, cold-blooded iguana turns food into protein. An iguana consumes less than 5 percent of what a chicken or cow eats to produce an equivalent pound of meat.

Instead of clearing the forest for pasture, researchers are urging farmers to raise animals that do well in the forest. They calculate that a farmer could raise 100 iguanas a year on a typical two-and-a-half-acre plot of forest. At an average 6½ pounds each, the iguanas would produce 650 pounds of low-fat meat, better than the average cattle ranch.

Dr. Dagmar Werner, director of the iguana project, said she tested the idea in two Panamanian villages by giving farmers cages, incubators, and expert advice on nurturing iguanas.

"Nobody burns the forest anymore," she told the *New York Times*. "The villagers are very enthusiastic and are planting trees like mad."

I wondered how long it would be before mesquite-grilled iguana appeared at the next ecologically hip reception.

THERE IS A CHANGE coming to the Amazon. These visionary explorations of ways of making a living from the rain forest are part of a future for the forest.

But the iguana ranches are still a long way from being a commercial success. And while marketing rain-forest products is a promising avenue, the hype that has surrounded this approach to "saving the rain forest" has obscured some real problems.

Dr. Werner provided too much help to her fledgling iguana farmers for her experience to reflect real-life conditions. And Ben and Jerry are having a hard time guaranteeing the quantity and quality of nuts needed for their high-test ice cream without going through the local Brazil-nut barons who have long exploited gatherers.

Besides, in the Brazilian Amazon where I traveled, trade in "extractive" products supports less than 13 percent of the population, and the sector's contribution to the regional economy has declined steadily since the last rubber boom. Close to 15 million people live in the region now, more than half of them in cities.

I thought about the people in those cities. I found hope in imagining that in the Amazon, like in the American West to which it has so often been compared, the facts of the landscape will still dominate human endeavors in the future. Here it is a desert, there a rain forest, but both environments exact careful human attention to the realities of the land and a heavy price from those who fail to pay attention.

The great Amazon Bason is a landscape changing on a human scale before the eyes of the world. Maybe iguana ranches will be part of that landscape in the future, along with extractive reserves, national forests, parks, and Indian reservations, but so will cities, dams, highways, cattle ranches, farms and mines. ■

## The Fate of the Forest

*This is the first of a whole forest of new books to look at the Amazon after the death of Chico Mendes, whose murder suddenly cast rubber tappers on center stage as the true guardians of the rain forest. After exploring what has made the Amazon such a hot topic lately — according to Hecht and Cockburn it is "the symbolic content of the dreams it ignites"* — **The Fate of the Forest** *lays out a clear, concise history of the political economy of the Amazon.*

*Popular myths of why the Amazon is going up in smoke are soundly debunked — it's not due to exporting cattle or timber, it's not because of multinationals, and it's not because of Brazil's huge debt. The authors show how the Brazilian military's dream of occupying the Amazon spawned a feeding frenzy of tax credits and incentives for deforestation for wealthy companies and individuals from the developed south of Brazil. But they take pains to clarify that fire is an essential part of tropical agriculture and that cattle can be a rational hedge against inflation for set-*

*tlers and forest dwellers in an economy where banks mean nothing and money means less every day.*

*Hecht and Cockburn also give a most accurate rendering of Brazil's grassroots struggle to save the Amazon, a fight more for justice than ecological preserves. They don't flinch at portraying Chico Mendes as the "extremely radical political militant" that he was as a rubber-tapper union organizer. It's a pity, though, that in their effort to reduce the hopes of the Amazon to "socialist ecology," the authors lean too heavily on a tiresome class analysis and, to my disappointment, invoke messianic Brazilian rebellions that ended in massacres in the past, as if they should be models for the fate of the forest. They leave me fearing for the new martyrs who will fall for the cause.* —Jon Christensen

•

Fire is essential to the management of humid tropical forests for human purposes. Only an American generation brought up on the ursine caveats of Smoky the Bear would find anything odd about this. . . .

The issue in the rainforest is not fire itself, but its purposes. Does its use in a particular instance inhibit regeneration, diminish the diversity of species and waste nutrients, or is it part of a process through which this diversity is enhanced, nutrients recaptured in new vegetation, and regeneration encouraged?

•

Land in the Amazon became a vehicle for capturing incentives, cheap credits, and itself assumed the form of a speculative instrument and an object of exchange rather than being an input into agriculture.

**The Fate of the Forest**
Alexander Cockburn and Susanna Hecht 1989; 288 pp.

**$9.95**
($13.45 postpaid) from HarperCollins Publishers/Customer Fulfillment, 1000 Keystone Industrial Park, Scranton, PA 18510; 800/331-3761 (or Whole Earth Access)

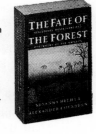

# Miracle in a

How could any place be so lifeless?, I wondered. Not a wisp of grass showed anywhere. Not an insect crept among the rocks. I was standing on fossil coral rubble left in an abandoned quarry near Mombasa on the coast of Kenya. It was a man-made desert, created after the Bamburi Cement Company had dug out 30 feet of limestone.

Yet a five-minute walk away, over a ridge on the same Bamburi site, I found myself in one of the most luxuriant scenes I have encountered in Kenya. Trees towered all around. Lush grass proliferated underfoot. Insects swarmed and birds flitted about. Up ahead, fish bunched in a series of pools. A hippo wallowed contentedly. Crocodiles sunned themselves. Beyond them, cattle, sheep and goats grazed with herds of semi-domesticated eland and oryx.

"It's surprising what nature can do when you lend a helping hand," explained my companion, Rene Haller, a stocky Swiss agronomist. "The trick is to get something started, a few plants, then let nature get on with it Africa-style." A self-trained specialist in restoring man-blighted lands, Haller, now in his 50s, is manager of Baobab Farm, a Bamburi subsidiary set up to reclaim the quarry. "The company asked me to do something, anything, to conceal the scars of its quarrying," he says. "Here you see some results of working with nature."

Haller's "results" are remarkable. He has restored more than 250 of almost 900 acres of blasted landscape, but the significance of his work goes far beyond the mere reclamation of a derelict quarry. Haller has made Bam-

## Quarry Restoration In Kenya

### BY NORMAN MYERS

*PHOTOGRAPHS BY WILLIAM CAMPBELL*

# Lifeless Pit

**(Left) Rene Haller has helped transform a Kenyan cement quarry into a food- and timber-producing nature preserve called Baobab Farm (far left).**

**(Below left) An oryx grazes on part of 258 restored acres, with the cement plant visible in the background.**

*Robert Horvitz pointed us towards this story, after hearing a version of it on the radio. Quarries have long been sites for restoration; Canada's Butchart Gardens in Victoria, British Columbia, is one of the oldest and most famous. Norman Myers is the author of* The Sinking Ark *and* The Primary Source. —*Richard Nilsen*

buri a ray of hope for a continent with some of the worst environmental problems on Earth. In the process, he has demonstrated to small-scale farmers that they, too, can make their marginal lands fruitful. If Haller's message can be spread widely enough, throngs of peasants, whose rudimentary agriculture now tends to lay waste to their soil, can regenerate their environments and enhance their living standards.

Haller's success has been to combine his own brand of low-cost pragmatism with time-proven customs and new farming techniques. Development experts have long proclaimed the virtues of using local plants and animals for innovative agriculture — especially for a technique known as agriforestry, which combines growing trees with raising crops and livestock. Trees supply fuelwood, building poles, foods and livestock browse, and some species add natural fertilizer to the soil. Above all, the theory goes, trees protect the African earth, and they represent the best approach to the regreening of the continent. The problem has been producing the goods without costly inputs of fuel and fertilizer. Yet Haller has managed to get it all

together using a return to traditional agriculture — and brains rather than money.

Meticulously organized and ever inventive, Rene Haller exudes effervescent good spirits. He is one part bright-eyed man of ideas, another part hard-nosed entrepreneur. "There's always a better way," he says, and "everything must pay its way."

Haller's African success story began in 1959 when he arrived in Mombasa to join the cement company. Bamburi was already a major Kenya enterprise, mining prehistoric beds of coral for limestone to manufacture cement. Cement was so precious in a developing country that at first nobody objected if the operation left horrendous gashes across the landscape. But the company began to grow sensitive about its scarred-earth image. So it turned to agronomist Haller. They would pay his salary to restore plant cover, give him some staff, and lend him a little equipment — that was it. Just the sort of challenge Haller welcomed.

But how to get anything, anything at all, to grow on bare coral with virtually no underground water? Walking with Haller around a newly abandoned patch of quarry, I scuffed the surface with my boot. The action left more impression on my foot than on the ground. "The earth is so rock hard," explained Haller, "that I have to get my staff to dig holes with picks before I plant trees."

All told, Haller has found only a half-dozen tree species that can prosper in such hostile conditions. The best one appears to be casuarina, a needle-leaved tree that soars once it gets a foothold in the harsh Bamburi terrain. It is one of those specialized plants that supply their own nitrogen fertilizer by plucking the raw material out of the atmosphere. So vigorously have Haller's casuarinas grown that his grove of several dozen acres produced no less than 50 tons of fuelwood after only five years of growth.

Two other species, planted in tandem, tell a special tale of Haller's ingenuity. The fast-growing *Conocarpus*, viewed by traditional foresters almost as a ''weed,'' has been scarcely used on plantations despite the excellent charcoal it produces. Its drawback is that it requires fertilizer. The *Prosopis*, a relative of the American mesquite shrub, is not so super-speedy at growing, but it fixes atmospheric nitrogen. It also produces masses

*Harvesting tilapia, a fast-growing fish, on Baobab Farm. The fish are sold on the local market.*

of flowers for honeymaking, plus fruits suitable for livestock feed. A plantation of both trees produces a far better forest than if the two species are grown separately.

This kind of interrelatedness is the key factor in all of Haller's activities. ''Nature does not do things by compartments, nor should we, if we are to take advantage of the full natural bounty of the Earth,'' he philosophizes.

His livestock program illustrates that principle as well. Baobab Farm's 258 acres support 500 sheep, 160 goats and 25 cattle. The animals feed on grass that now grows among the trees in a thin layer of new soil produced from leaf litter. They also eat foliage and other sorts of fodder generated by the diverse types of plantings. ''If my pasture were open grassland, the livestock would not survive a drought year,'' Haller says. ''But the forest provides shade as well as food, and it saves them during the drier periods.''

In among the domestic animals roam 25 eland and 20 oryx, antelopes that normally live wild in Kenya. They are hardy and need next to no water. They also eat different plants from those consumed by conventional livestock, so they do not compete for scarce forage. They produce more meat per unit of animal weight than does a typical African cow, contain more protein, and if Kenya's laws banning sale of wildlife products are changed, could yield commercial meat.

Still another source of protein produced at Baobab is fish. Initially, Haller used the forest lakes at Bamburi to experiment with aquaculture, stocking his ponds with tilapia, a succulent fish that proliferates in East Africa's natural lakes. Since the ponds contained weeds and algae, he left the tilapia largely to their own devices. Then, following the lead of experts involved in intensive fish-farming in Asia, he built a series of fish-rearing tanks, restoring the ponds to natural pools once again.

The outcome is an unparalleled success. As Haller enthusiastically explains, raising domestic fish is probably the most efficient method known for producing animal protein. The fish need just under two pounds of feed to produce one pound of flesh. By comparison, cattle generally have a ''conversion ratio'' of about 10:1. Haller's intensive system produces 35 tons of fish a year from a mere one-third of an acre. It is at least as economic, he points out, as going to sea in a boat to catch ''wild'' fish.

As with most of Haller's ideas, there are easy-to-duplicate lessons for local people. Every ''Shamba,'' or smallholder plot of a few acres, has space for a simple pond lined with clay, gravel or sand. The owner throws in a few fingerlings, then occasional handfuls of farm garbage. And

*Transplanting tree seedlings in a quarry nursery. Once established, the decomposed leaf litter from these trees will create scarce topsoil. The bag-like plastic containers are cheap and easy for workers to carry.*

that's it. A 10-square-yard pond can generate more animal protein in a week than many Africans get in a month.

Already, Haller's package has been adopted throughout much of Kenya and in several other countries of Africa, principally by farmers' associations, village cooperatives, and government and development agencies. In addition, his methods have won the praise of the United Nations agency that leads the field in fish affairs, the Food and Agriculture Organization. He's also attracting the interest of the World Bank, an agency that normally deals only in superscale projects.

Meanwhile, the start-out ponds have steadily attracted an array of birdlife — 130 species in all. The pond area and the original forest expanse have been fenced off, and the 18-acre enclosure now accommodates several buffalo, bushbuck, duiker, suni, warthogs, hippos, civet cats, mongooses, snakes, and crocodiles. Haller now charges admission to tourists to see this new side of his operation.

Local people, who also visit, are introduced to other aspects of Haller's integrated philosophy. One of the newest is crocodile farming. Haller feeds 700 young crocs a diet of fish offal and waste from livestock carcasses. The crocs, raised in the original fish ponds and in a crocodile compound, cost nothing to keep, and when they reach sufficient length within just a few years, Haller will make a tidy profit by marketing their high-value skins. In turn, their meat will be fed back to the fish.

And that's only the beginning. In moister parts of his farm, Haller harvests surplus earthworms which offer top-grade protein for his tilapias. In waste-flow channels, he raises giant African snails for human food. And he plans to use fish excrement and general fish-pond sludge to generate biogas for energy. He has introduced millipedes into his casuarina groves to speed the transformation of leaf litter into soil. He is growing grapes for winemaking on steep hillsides, collecting honey from hives he established in his *Prosopis* trees, and even growing bananas using ingenious but simple methods to coax moisture from the normally dry soil.

Such inventiveness may be an antidote to many of Kenya's problems. The bulk of the country's population is made up of small-scale cultivators — and the population, now 20 million, is already overloading the nation's capacity to support it. For these hard-pressed multitudes, whose numbers may double by the year 2000, the value in Haller's operation is that a farmer can often get the most from his patch of land by enabling it to produce food, fiber, and fuel all at once in a mutually supportive mixture. In the past, Western agriculturalists have generally taught Africans that segmented plots and monocultures are "good farming" and that a mixed system is messy. Now those ideas have to be unlearned.

Part of the beauty of Haller's model is its low cost. Baobab Farm is essentially self-financing. Although Haller is on the company's payroll, he supplies fish, meat, fruit, and fuelwood for the company's work force, and he sells more to outsiders. He also gets revenues from those organizations that pay for his expertise, and he collects entrance fees from tourists. It is a measure of the profitability of his enterprise that it keeps on expanding — and that Haller can keep on trying out new ideas. "I would not want anything else," he explains. "How could I possibly promote it to others if it did not stand up commercially?"

And Haller's ideas are being promoted. As word gets around, more local farmers are coming to see what they can try out back home. I met such a group, from near Nairobi 300 miles inland. Why had they traveled all that way?, I asked. "Well, our children are many and our acres are few. The chief of our village tells us we must look for fresh horizons in here," their leader said, pointing to his head. "We had heard of this Swiss man who smiles and makes something out of nothing. We want to see for ourselves."

Similarly, I talked with a delegation from farms 50 miles away along the coast, on their way out from the quarry forest. "Now we see the difference," one of them said, "between wasteland and wasted land. Here is a forest growing on a desert. And I heard that some trees are fertilizer factories, but I thought it was just another expert's tale. My eyes tell me the truth." ■

*Sally the hippo with another revenue source for Baobab Farm — tourists. A wild-born orphaned animal, Sally is semi-tame. She was raised by a maker of wildlife films, but moved here after she wandered onto his porch and ate his couch.*

# Come Back Salmon

## A Japanese Environmental Success

*BY TOM MURDOCH*

**Salmon are returning to Japan. This boy is about to release a fingerling into the Tamagawa River.**

**A poster made by elementary-school kids. "SOS" means Save Our Salmon.**

*Tom Murdoch began the Adopt-A-Stream Foundation in Washington State in 1981 (for more information, see page 123). Here he reports on a sister organization. Keeping streams fit for fish — and getting schoolchildren to take on that responsibility — is an idea that has also caught on in Japan. Pay attention to a common species like salmon and the Pacific Rim begins to shrink.*
*—Richard Nilsen*

Cries of "Come Back Salmon" are now being heard on the rivers of Japan. Motivated by Japan's Come Back Salmon Society (CBSS), schoolchildren and their parents and teachers are raising and releasing baby salmon from Tokyo north to the rivers of Hokkaido (the large northern island where the city of Sapporo is located). Salmon associations now exist on every river north of Tokyo, and more than 240 Japanese elementary and high schools have developed environmental education programs focusing on the revered salmon resource.

How did all of this happen?

The CBSS was privately initiated in the late seventies by Hokkaido University anthropology professor Masakazu Yoshizaki and a small group of influential residents of Sapporo. This group established the difficult goal of getting salmon back into the Toyohiro River, which flows through the middle of downtown Sapporo.

The Toyohiro River had lost its salmon run more than 25 years before, due to a variety of factors including rapid development and industrialization with little regard to environmental protection. In 1978 the Come Back Salmon campaign got thousands of schoolchildren planting baby salmon in the Toyohiro, removing tons of garbage and debris from the river, and waking up the adult population to pollution problems.

In 1982 I had the good fortune of being invited to Hokkaido to share environmental education and stream enhancement information with the CBSS, Canada's Save the Salmon Society and Salmonid Enhancement Program, and the UK's Thames Water Authority. I witnessed the first returns of salmon to the Toyohiro River in decades.

Several highlights of that trip stuck with me:

*Private support:* Takara Shozu LTD, a major distillery, donated 2 percent of its profits to the CBSS for five years to promote environmental education.

*Direct media involvement:* The news director for Hokkaido Television and Broadcasting, Anshin Sugawara, was also the Secretary General of the Sapporo Salmon Society. Sugawara, often referred to as "Uncle Salmon" in Sapporo, raised salmon in the TV studio and reported on their growth and progress every night. He invited viewers to help release them, and 5,000 people showed up.

*Public environmental education centers:* A Shinto shrine in the center of Sapporo, which had a spring-fed pond, provided local schools with salmon-rearing facilities. A major Salmon Museum is now completed near a city park — a wonderful education facility. There are salmon associations on all of

Hokkaido's rivers. And salmon now return every year to the Toyohiro River.

*School involvement:* All the elementary schools in Sapporo had environmental education programs that seemed to involve every academic discipline. I am the proud possessor of more than fifty beautiful wood-block prints focusing on

Japanese schoolchildren visit the United States and learn how to monitor water quality. Salmon need clean water in order to breed. Here an American student holds up a water sample taken from Pigeon Creek near Everett, Washington.

the life cycle of salmon, created by seventh-graders.

Salmon are beginning to run again in Tokyo's Tamagawa River. Picking up on the Sapporo model, the Tamagawa River Salmon Society stimulated 60 Tokyo schools to raise and release baby salmon from a jointly operated mini-hatchery and classroom aquariums.

"We try to convince the youngsters that when the baby salmon are released into the river, they then must do everything possible to make sure that the river will be a suitable place for these babies to return to and spawn future generations," says Rensei Baba, the society's business manager. Mr. Baba is also the science writer for the *Yomiuri Shimbun*, which has a daily readership of more than 9.5 million. Media support for this small environmental organization has been crucial, and is growing.

Last February, the *Yomiuri Shimbun* and the Setagaya Ward of Tokyo jointly sponsored a conference with the theme of "getting children involved in the environment." The stars of this event were speakers from the US, Canada, England, Austria, the Soviet Union, and (of course) Japan . . . most of whom were fifth- and sixth-grade students.

I was able to witness this first governmental involvement and support for Come Back Salmon's efforts. My rambunctious students from Everett, Washington, suggested to their Japanese counterparts that they start visiting politicians to request much more support. Perhaps this small environmental light in Japan will begin to glow much brighter in the future. ■

A salmon release ceremony on the banks of the Tamagawa River in Tokyo.

## Man of the Trees

*Richard St. Barbe Baker was a man ahead of his time, so far ahead in fact that nearly a decade after his death, the reforestation concepts he was preaching sixty years ago are just beginning to enter the popular consciousness. He traveled the globe teaching people the importance of planting trees, and his description of forests as the lungs of the earth anticipated in a conceptual way the Gaia Theory debated today. It's fair to call him a pioneer of environmental restoration.*

*The book* **Man of the Trees** *is short selections of St. Barbe Baker's writing, taken from seven of his many books. The video by the same name is a documentary of his life, and includes a great sequence with him hugging a big tree as*

**St. Barbe with a planting team of Beijing schoolchildren in front of the Chinese Academy of Forestry in 1981.**

*he explains that he does this daily as a way of gathering energy and charging up his batteries. Who knows, in another sixty years maybe we'll understand that too.*          —Richard Nilsen

•

Today it is the duty of every thinking being to live and to serve not only his own day and generation, but also generations unborn, by helping to restore and maintain the green glory of the forests of the earth.

**Man of the Trees**
Video (VHS)

**$23** postpaid; catalog **free** from Music for Little People, P. O. Box 1460, Redway, CA 95560; 800/346-4445.

**Man of the Trees**
(Selected Writings of Richard St. Barbe Baker)
Karen Gridley, Editor
1989; 115 pp.

**$12.50** ($13.50 postpaid) from Ecology Action, 5798 Ridgewood Road, Willits, CA 95490.

### The Man Who Planted Trees
Video (VHS)

**$33.45** postpaid from CBC Home Video, P. O. Box 6440/Station A, Montreal, QC H3C 3L4, Canada; 514/597-4040.

## The Man Who Planted Trees

*Jean Giono's short story about a tree-planting hermit in the south of France is a parable of restoration. The oaks and beeches sown by this one man over forty years transform an entire region which had been degraded by deforestation. The animated film of this inspiring story is not just good, it's excellent — winning an Oscar in 1987. Frederic Back's drawing is soft, shimmering, and uses a point of view that's constantly in motion. Christopher Plummer narrates. The video version is packaged with two other animated shorts by Back; one of them,* **Crac!**, *also won an Oscar in 1982.* —Richard Nilsen
[Suggested by Tom Waller]

## Backyard Safari

*Jack Schmidling and Marilyn Schenk show how they created a mini-prairie in their Illinois backyard. First the how-to part: build a fence so neighbors won't complain about your "unkempt" yard; plant native plants (milkweed and thistle — weeds to us — are manna in the urbanized desert to goldfinches and monarch butterfly larvae); burn the "prairie" in early spring so grass and flower seeds will germinate. Then the safari begins — a trip through a yearly cycle in the backyard prairie ecosystem. Here the wildlife (beautifully filmed) show the payoff for converting your backyard from barbecue patio to wildlife sanctuary. From lacewing beetle to the rabbit that visits this postage-stamp paradise, the diversity of life in a small backyard is astounding, entertaining, humbling. Don't be surprised when this video ends with a plug to join the Backyard Wildlife Sanctu-*ary Program (sponsored by the National Wildlife Federation) — a good, do-able project for homeowners everywhere.
          —Corinne Cullen Hawkins

### Backyard Safari
Video (VHS)

**$32** postpaid; catalog **free** from Jack Schmidling Productions, 4501 Moody, Chicago, IL 60630; 312/685-1878.

# 4.

# PRACTICE OF ENVIRONMENTAL RESTORATION

S O HOW DO WE GO about repairing what has already been mucked up? For many of us, the first step is to slow down enough to observe the natural processes already at work, and to realize that restoration goes on for more than just our lifetime. It is both a big process and very discrete, and it begins with our actions in specific places. Here are techniques, tools and information sources to help recognize those places and know how to act responsibly. Healthy environments have many components, and maintaining them can involve more than environmental restoration.

A look at some of these other tools concludes this section. Sanford Lewis and Philip Wexler share new ways to address the threat of toxic pollution. Responding to the chemical soup of industrial society is often a necessary precursor to restoration work. It's a little like discovering an overflowing bathtub — before mopping the floor you first turn off the faucet. Here are approaches that citizens are learning to do for themselves, and the new information tools and political strategies that are making them work. There are also new tools coming to the aid of endangered animal species. Judith Goldsmith describes an Oregon laboratory that is helping the survival chances of endangered animals around the world. Hopefully these stories of people figuring out how to do effective environmental restoration will help you to write some of your own. Good luck!

Jack Monschke

# How To Heal The Land

### BY JACK MONSCHKE

I HAVE CONSCIOUSLY SHIED AWAY from writing about doing restoration work for years for two apparently contradictory reasons. First, it all seems so simple and obvious to me: water flows downhill. The other side of this coin is that each erosion problem and the resulting attempt at restoration is totally unique and site-specific. It's not clear-cut. I depend, not on a scientific formula, but on getting the feel of a situation. How can I write about that?

I am hoping here to set out a way for you to get a feel for the land, a way of looking at a situation so that you really see it, and some general guidelines that will lead you to discovering for yourself how to lessen our human impact on the land in general and how to heal damaged land. When I am beginning a project, I spend a lot of time just walking the land and quieting myself. Then I begin to take notes, and the ideas just come. I have incredible respect for the power of nature and how little control I

*Every piece of land is different. This perception has even entered our language in the term "site-specific." Every restoration project is unique as well. The principles behind the work remain the same — careful observation, timely intervention, willingness to experiment and to stick around for the long haul.*

*The land in this article is 290 steep, logged-over acres near the headwaters of Salmon Creek, a tributary of the Eel River, on California's north coast. Jack Monschke, 45, has been restoring it since 1972. He is currently a consultant for a timber company, dealing with logging practices and long-range watershed management. Jack's wife Jonelle provided the illustrations.*
—Richard Nilsen

Jack Monschke

(Far left) Jack and nine-year-old Kasara stand in exactly the same place they stood eight years ago (center), when the same section of the creek had just undergone heavy-equipment work. Today the creek bed is restored, and planted trees prevent erosion. In 1981, before any work was done on it, the creek bed was filled with debris (upper right).

have over it. Yet, I was recently inspired by this statement in a SEVA newsletter: "The worst mistake is to do nothing because you can only do a little." —Edmund Burke.

The great thing about doing restoration work is that you get to work with the incredible healing potential of Mother Nature. When you recognize this natural process, problems that seemed hopeless can be solved and you can accomplish amazing things. The first step in healing is to get to know the land, walking and observing it through all seasons, learning to understand the process that *is* healing, and then carefully and respectfully nurturing this natural process.

A specific way to get to know the land is in terms of watershed. When a drop of water hits the peak of your roof, it goes down one way or the other, and runs down to the valleys where it is concentrated. Just as the valleys in your roof drain the slopes of your roof, so the valleys in nature drain the slopes from the ridgetop on one side to the ridgetop on the other. The entire area that is drained *is* the watershed.

So the first thing you can do is identify the watersheds on your land, and I'm talking about really small watercourses here because that's how this all begins. Begin by following running water upstream (often in small drainages water will be flowing only when it's raining). By following the watercourses upstream you will eventually be led to the divide between watersheds. This can be almost flat, or jagged and steep (think of the Continen-

tal Divide). If you watch the water carefully, you will eventually find the place where it runs one way on one side and the other way on the other side. Identifying and understanding these watersheds large and small is a way to see what's happening on your land and a key to learning how to make changes where changes need to be made.

Erosion of the soil takes place where the binding and protective qualities of the earth are insufficient to stand against the energy of moving water. Much current erosion is the result of man artificially altering the watersheds, either by changing the vegetation, increasing the volume of water, or changing the gradient, all of which are interdependent. Water has great energy, and this energy increases when it is concentrated and when it drops fast. You can see when erosion is happening by noticing whether water is running clear or muddy. Muddy water means that active cutting is happening somewhere in that watershed. In the very simplest terms you can alleviate erosion by dissipating the energy of water at the point where it meets the soil.

If you build a new house that doesn't have rain gutters, the water drops off your eaves and begins cutting into the bare soil below. Things have been altered by your building: the gradient has been changed because the slope of your roof is different from the slope of the land that it covers, and the vegetation has been eliminated. The energy of the water has been changed, and the protective covering of the earth has been removed. You can mitigate this erosion in several ways. You can plant something below the eaves that will break the impact and hold the soil *or* you can physically protect the soil by covering it with a deck or gravel path *or* you can eliminate the water by placing a gutter. However, if you do put in a gutter, you are further concentrating the water and must therefore intensify your protection at the point where you bring that water down.

Say you are out in the rain following your watersheds around, and you see a rivulet or stream that's muddy. Follow it upstream. If you come to a fork, you might see a clear stream and a muddy stream running into each other. Keep on the trail of the muddy one. Sooner or later you'll come to the place where the erosion is actually occurring. If you're lucky, there will be a single, easily identifiable source: your dog just dug up a gopher hole, or a small bank is caving in because a branch diverted water against it, or a muddy rivulet is running off a poorly drained dirt road. Sometimes, though, there will be a number of problems, some of which are less easy to identify, and it's the accumulation of many conditions — the cumulative impact — that is turning the water muddy.

Splash (or sheet) erosion is caused by raindrops hitting bare mineral soil. Soil should always have a protective cover, either living or dead, i.e., plants or mulch and preferably both. The method I use most often to protect bare soil is to seed and mulch.

I use different seed mixes depending on the site (steepness, soil type, soil moisture, etc.) and its use (pasture, future forest, front yard, etc.). A mix that I have found adaptable for many different conditions is 3/4 annual rye and 1/4 perennial rye. I often add clovers and/or other legumes. (A good source for more information for your specific area is your local nursery, an agricultural extension office, or Federal Soil Stabilization office.) I then cover the seed with a layer of straw. Approximately 4,000 lbs. per acre, or 10 lbs. per 100 square feet, works out well because it allows the seed to come through while still providing physical protection from splash erosion. There is a delicate balance between the protection of a mulch and the growth of a living ground cover, and the choice of the correct combination is very site-specific. The mulch doesn't have to be straw; any sort of organic debris, like branches or leaves, will do. For a road, use gravel or pavement. The bottom line is that bare earth needs protection or it will erode. If it's an emergency situation, even plastic can be used to cover the soil. The ideal time to seed and mulch in the Pacific Northwest is in the fall just before the onset of the winter rains.

Another way erosion commonly occurs is through bank cutting — when the force of the water flowing in a channel cuts into the side(s) of the channel because of a physical diversion (small slide, uprooted tree, etc.), or because of changes in the gradient,

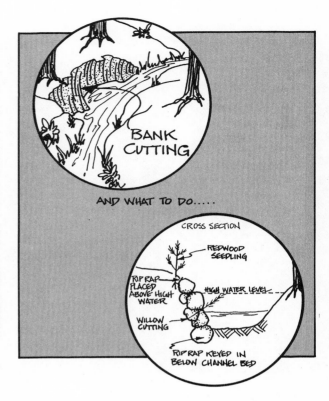

AND WHAT TO DO.....

the volume of water flowing the channel, or the vegetation on the banks. My first choice for protecting an eroding bank is to line it with rock; this is called riprap. You can also use wood, sod, or plantings of heavy-rooted plants that don't mind having wet feet. Even though rock riprap provides the best immediate protection for the bank, I always try to include plantings within the riprap for stability, appearance, and the long-term benefits to the ecosystem. There are certain trees that thrive along streams. They don't mind having their roots under water or having a few feet of silt dumped on them, and if they fall over, they have the ability to resprout. In our area redwood, alder, and willow have these characteristics, and I use them a lot.

Erosion also occurs from downcutting. This is where a gully is forming and getting deeper; think of the Grand

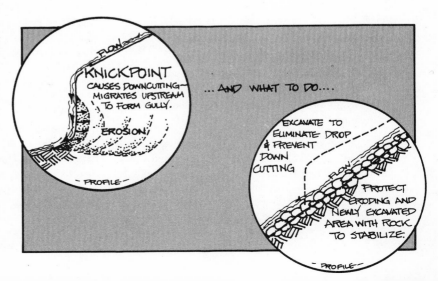

Canyon! If this is happening, you need to look for the knickpoint, which is the point where there is a dramatic change in the gradient of the watercourse. There are many different reasons why a knickpoint develops. It can be caused by something quite small like a rock or a branch on a small watercourse. Because of the nature of kinetic energy, the cutting power of the water increases geometrically with the increased drop and eventually creates a gully, which migrates upstream. First, you want to stop the downcutting at the knickpoint by eliminating the drop. Rock is best to use here because it not only holds firm, but also because it dissipates the force of the water.

Think of a waterfall. The cutting force of that water as it drops is incredibly strong. But in a cascade, where the water flows along from boulder to boulder with lots of aerated whitewater, the force of the water has been broken up. Wherever you see white foamy water, you know that its cutting force has been reduced. By changing the knickpoint from a fall to a cascade, you greatly reduce the cutting energy, and if rock is placed correctly, the gully will stop its migration upslope. If the bottom of a gully has no protection, it can be actively eroding even without a specific knickpoint. Rock, woody debris, heavily rooted grasses can all help stop active downcutting when placed in the bottom of a gully.

If you put something in the bottom of a freshly cut V-shaped gully, however, it serves to transfer the force of the water away from the bottom of the gully to the sides,

Jack Monschke

Redwood-slab check dams dissipate the force of the water, creating a graceful, gentle fall.

CHECK DAMS

CHECK DAMS "CHECK" OR DISSIPATE THE ENERGY OF THE FLOW & THUS PROTECT THE CHANNEL FROM EROSION.

VIEW A

VIEW B (PROFILE)

NOTE: THE STEEPER THE SLOPE, THE CLOSER TOGETHER CHECK DAMS MUST BE. A POORLY PLACED CHECK DAM CAN CREATE A KNICKPOINT & THUS A GULLY.

VIEW C

CHECK DAMS CAN BE MADE OF ROCK, WOODY DEBRIS, REDWOOD BOARDS, OR PLANTS. VIEW C SHOWS PLACEMENT OF JUNCUS OR SEDGE IN PAIRS. FUTURE GROWTH CREATES ENERGY DISSIPATING CHECK DAMS.

and then you might have bank-cutting. So you often need to protect the sides as well, by lining the channel in such a way that both the bottom and the sides are protected. If material and money are limited, you can build a series of check dams, which means you don't try to line the whole channel with rock, but do it intermittently. Check dams can accumulate sediment and dissipate the force of the water effectively if they are constructed properly. But check dams can be tricky to build, are site-specific, and are never as trustworthy as lining the entire channel. If experimenting with check dams is appealing to you, start slowly on a small, non-critical drainage and then watch your results carefully. This way you will get the feel of how they work in your specific situation.

I've found that juncus planting works as effectively on meadow gullies as living check dams. (Juncus is a heavy rush natural to this area, and it is very tolerant of wet soils.) As the years pass, the juncus grows, traps sediment, and turns the V-gully into a U-gully and finally into a swale. In fact, you can throw juncus and sedge roots into the bottom of a gully before placing rock or woody de-

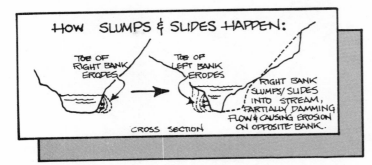

HOW SLUMPS & SLIDES HAPPEN:

TOE OF RIGHT BANK ERODES

TOE OF LEFT BANK ERODES

RIGHT BANK SLUMPS/SLIDES INTO STREAM, PARTIALLY DAMMING FLOW & CAUSING EROSION ON OPPOSITE BANK.

CROSS SECTION

bris, and they will often re-sprout through the rock the next winter.

When you have bank cutting occurring along a stream, at some point the cutting may become so great that the toe of the sloped bank is removed and the undercut bank slumps into the stream. This brings us to what I call slumps and slides. This action can set up a chain reaction. The slide creates a temporary dam which leads to bank cutting on the other side of the watercourse, which creates a slide, and so on. Protect the bank, and especially the toe of that bank, with something — rock, if possible. Try to keep the water off the toe. With some slumps and slides, the water which is supersaturating the soil and contributing to its instability is actually subsurface. Then you need to put in drainage facilities to release that water to the surface. You might also find that water is running onto that sliding toe from above, and in that case you want to use waterbars or diversions to keep the water off that unstable section of the bank. Sometimes, in a very difficult situation — and I have this exact situation on my road — all three of these conditions are present.

If you have a situation where big slumps and/or slides are

happening, first spend some time just watching it and try to get a feeling for what's going on. Then call for help. Finding experts in this field can be difficult depending on where you live and what your problem is, but for starters find the most experienced and skilled heavy-equipment operator in your area and then determine whether that person has experience dealing with similar problems. Seeking help from professional experts — engineers, geologists, hydrologists, etc. — can be cost-prohibitive for the small landowner, and it's often possible to get very sound advice from the person who will actually be doing the work.

When I first started creek restoration work, I tried to do it all by hand. I quickly learned that it wasn't practical to go in and remove a few hundred yards of fill by hand. Heavy equipment is often necessary, although after the

Jack Monschke

**(Before)**
Section of creek where bank cutting was causing a slide. The bank was an unhealed wound constantly being eroded by high flows. The shallow creek bed was a poor habitat for fish.

**(After)**
Tree planting and rock protection on the eroding bank have stopped all active erosion. The pool here is approximately 30' in diameter, an excellent fish habitat.

heavy equipment work is done, I almost always go back in and touch up the work by hand, placing small rock, etc. I am very lucky in that I work with a highly skilled heavy-equipment operator. We can go into a creek, and he will not damage any vegetation we have chosen to keep. But your average random cat skinner is not this skilled or sensitive, so you need to be selective when choosing one.

Sometimes watercourses are diverted totally out of their natural channels. In the Pacific Northwest on the vast acres of logged-over lands, these diverted watercourses are often the single most common source of stream sediments. When the loggers arrived in the old days, they cut roads wherever they wanted, and often when they were done they left the roads in place without putting

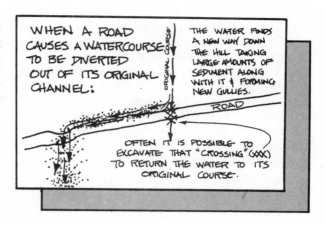

WHEN A ROAD CAUSES A WATERCOURSE TO BE DIVERTED OUT OF ITS ORIGINAL CHANNEL:

THE WATER FINDS A NEW WAY DOWN THE HILL TAKING LARGE AMOUNTS OF SEDIMENT ALONG WITH IT & FORMING NEW GULLIES.

ROAD

OFTEN IT IS POSSIBLE TO EXCAVATE THAT "CROSSING" (XXX) TO RETURN THE WATER TO ITS ORIGINAL COURSE.

THE PURPOSE OF A WATERBAR IS TO GET THE WATER OFF THE ROAD WHEN IT RAINS...

ROAD

MAKE SURE THE OUTFLOW GOES INTO VEGETATION OR ROCK — NOT ONTO BARE ROAD FILL.

THE FREQUENCY & SIZE OF WATERBARS DEPEND ON MANY FACTORS: HOW STEEP YOUR ROAD IS, HOW WELL IT IS CROWNED (OR HOW SEVERELY RUTTED), HOW OFTEN YOU DRIVE IT (AND WHETHER YOU CAN STAND BIG BUMPS), ETC.

in any drainage facilities (culverts, bridges, waterbars, etc.). So many, many watercourses were diverted completely out of their natural drainages and forced to find new ways. This is a common problem, but it is often relatively easy to resolve: look for the original drainage and return the stream to its old home where it belongs. In some situations, however, the water has flowed in its newly formed gully for so long that to change it back may cause more erosion than it stops. This is a site-specific situation, and I advise careful observation over time. When the water is returned to its original channel, the newly formed gully still exists and may need to be treated.

Around here most of us live on land that was roaded with very little thought given to erosion, and roads are the main cause of most of the active erosion that's going on today. If you have the opportunity to build your own road, the most important thing to look for is natural watershed patterns. Identify these and then change them as little as possible. The best roads are built on ridges (those divides between watersheds) and are crowned so the water can run off both sides. A ridgetop road is the safest and most stable because when you build this way, you are not changing the gradient or the volume of water flowing in the drainages. You are only changing the vegetation, and

building narrow roads and leaving as many trees as possible will even minimize that.

A road on a hillside is constructed by cutting and filling. To lessen the impact, outslope the road gently (it doesn't even need to be noticeable) so that the water that comes off the slope and the water that falls on the road itself will flow across the road and down the hillside. Outsloping the road can be a safety problem if you live in an area with snow and ice because if you go into a skid, the tendency will be to slide to the outside of the road and over the bank. If outsloping isn't practical, an inboard ditch and cross-road drains must be installed.

Plant and mulch both the cut and fill, the upslope and downslope, to minimize the overall change to the vegetation and eliminate or minimize splash (sheet) erosion.

Keep all watercourses that the road crosses in their natural drainages. This is done by proper position and placement of culverts, a major subject in itself.

Steel culverts are expensive, and what happened around here was that the people who put in the roads tried to save money by putting in as few culverts as possible, with inboard ditches along the road. The idea was that the water from a little watershed would hit the road and flow

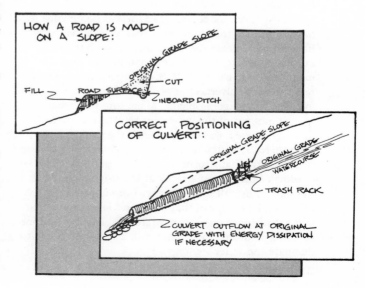

HOW A ROAD IS MADE ON A SLOPE:

ORIGINAL GRADE SLOPE

FILL

ROAD SURFACE

CUT

INBOARD DITCH

CORRECT POSITIONING OF CULVERT:

ORIGINAL GRADE SLOPE

ORIGINAL GRADE

WATERCOURSE

TRASH RACK

CULVERT OUTFLOW AT ORIGINAL GRADE WITH ENERGY DISSIPATION IF NECESSARY into and along an inboard ditch until it came to a culvert, at which point the water would flow under the road and get dumped out on the downhill side of the road. This doesn't work because the volume of water dumped out at a given spot gets greatly increased, the gradients are drastically changed, and the vegetation has been changed by cutting and filling to build the road in the first place. The results of this method are road failures, and bank failures both above and below the road, plus lots of eroding soil.

The right solution is to put in a culvert at every place where the road crosses a watercourse. They may be small culverts, but they are necessary so that volume and gradient of the runoff is altered as little as possible. The money spent on culverts when building a road right will save you a great deal of money on maintenance cost and also eliminate anxiety and eroding soil in the future. Culverts generally need to be placed by a backhoe or excavator, and it's important to check around in your area for a good heavy equipment operator, one with sensitivity to the land and water and the way they work together. A good backhoe operator will place your culverts correctly at the original grade and at the exact location of the watercourses it is draining.

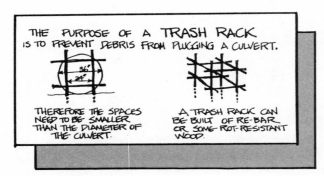

THE PURPOSE OF A TRASH RACK IS TO PREVENT DEBRIS FROM PLUGGING A CULVERT.

THEREFORE THE SPACES NEED TO BE SMALLER THAN THE DIAMETER OF THE CULVERT.

A TRASH RACK CAN BE BUILT OF RE-BAR OR SOME ROT-RESISTANT WOOD.

Make sure that as the water flows into the culvert on the inboard side, it does not flow under the culvert or erode the fill on the sides of the culvert because this will eat away little caves under your road surface. Sometimes you have to protect this point with rock or sandbag riprap. In order to decrease the possibility of a culvert plugging during winter storms, I recommend placing a trash rack made of steel re-bar at the inlet. A properly designed trash rack will keep culverts clear of obstructions without causing other drainage problems. Make sure that where the water flows out of the culvert, it returns exactly to that drainage where it was flowing before the road and culvert were there. This drainage has spent centuries adapting itself to carry that stream. If it is impossible to place the outflow of the culvert exactly at the original grade and the water must fall some distance onto unprotected soil, dissipate the energy of that water and lessen the impact of the gradient change you have created. This can be done by using a culvert downchute with a rock pile below to dissipate the water's energy.

Winter road maintenance is crucial. I walk my roads during the first good rain, checking all trash racks and/or culvert inlets to make sure there are no obstructions. I also check all ditches, both inboard and cross-road, to make sure they are capable of carrying the runoff of a heavy storm. The small ditches that drain the surface water off the road usually have to be redone every year, because summer driving wears them down. During a rain the water in these ditches should run fast enough so that it doesn't deposit sediment and plug, but slow enough to keep from downcutting. This is another case where close observation and fine-tuning is critical.

I also walk my roads during the first very heavy storm, removing debris if necessary and checking all culvert outlets to be sure there isn't any downcutting. Always walk your roads during any heavy flooding storm and follow the above procedures. Winter maintenance can save you thousands of dollars. It's preventive maintenance at its best.

The same rules that apply to siting a road also apply to siting a house and garden. Put your home and garden where they change the natural watershed as little as possible. Don't just bulldoze in a big flat and then have it slump away. Fill is very unstable. Unless you're willing to terrace it, a garden should not be on a grade of more than 10 percent. A garden is in constant disturbance, and in terms of erosion it never has a chance to heal. I think the most important thing in planning where to build and garden is to take the time to really get to know your land before making any changes.

One of the most beneficial things people can do for any land is to plant trees, which help prevent erosion in many ways. First, the impact of rain is broken up by the canopy; second, the roots and humus absorb much more water than bare soil; and finally, roots provide structural binding strength to soil. Although grass does this work also, it is less effective than trees. The water that has been absorbed by the ground is released slowly over time. In a healthy forest a lot of the water absorbed during the winter rains is released over the whole summer. I recommend planting trees along all roads and streams, and anywhere trees have grown in the past and have not reestablished themselves naturally. If you have a lot of non-forested land, and it used to support trees (stumps are the obvious indicator here), the land will support trees again. You might have to give the seedlings a little shade and water for a year or two, but getting the root structure back and the canopy overhead is one of the simplest, most effective and most rewarding things you can do to heal the land.

It's important to choose the right planting stock. The Agricultural Extension office and State Nursery office can help you get the proper stock for your area. I have had a much higher survival rate planting bare-root stock than containerized stock on harsher sites, and for that reason, I prefer to plant bare-root, although the planting takes more skill and patience. It was really hard for me to accept this, but trees like to be planted in bare, mineral soil. Bare-root trees don't like humus, and young trees are hurt

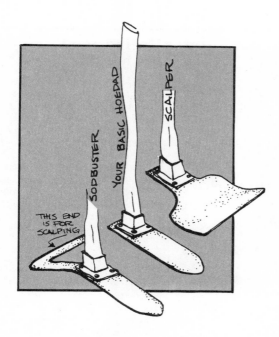

the native species. For example, I've never used Argentine pampas grass, although its root binding is great for erosion control, because it can take over a clearcut or other disturbed area, and then the native species can't get started. Sometimes I'll plant a non-native species as a nursery crop to slightly alter the micro climate and thus enable a native species to survive in the future.

If deer browsing is a problem, I have found that 3'' x 24'' seedling protection tubes have been very effective on young fir trees to get them above browse height. These tubes are stiff plastic mesh and protect the leaders of the little trees as they grow. You have to move them up just before new growth each spring. On fast-growing trees the tubes will allow the trees to get above browse height in 2-3 years, compared to trees I have watched that were stunted for 10-20 years without any browse protection.

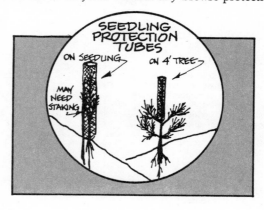

by competition with grasses. To prepare the ground, scrape off all the existing grass and debris down to the bare soil. This is called scalping. If you're working by hand, scrape off a 30'' square and plant in the middle of it. There is a great tool made now for scalping called, of course, a scalper. The tool for planting bare-root seedlings is called a hoe-dad. Both functions of scalping and planting are combined in one tool called a sodbuster, which is perfect for the smaller landowner who doesn't have large acreage to prepare for planting.

If you're planting a large area, have a Caterpillar do what is called mechanical site preparation. When I do this, I get the best tree seedling survival rate by clearing 10- to 12-foot swaths with the Cat's blade tipped so that all the grass and debris is pushed to the downside to make a berm. Then I leave a 10-foot section on either side of that swath untouched to act as a filter strip. It's important that these are made on the contour of the slope so that they act as terraces breaking up the energy and dispersing the concentration of runoff.

In California you can buy 1,000 one- or two-year-old tree seedlings for about $150 from the state nursery. If you plant them right, they will cover approximately three acres. There are government programs to help you reforest your land. Some people are suspicious of government programs, but I've had a very positive working relationship with government cost-sharing programs for 18 years. They've made it possible for me and my clients to do a lot of conservation work that otherwise would have been impossible because of cost constraints. Even though the programs are there to encourage forest products for future harvest, there's nothing in the programs that says you have to harvest.

There are times when the microclimate and the soil conditions have been changed so much that the natural vegetation won't survive. When I can't get a native species to work, I will introduce a non-native one, but I will check it out to see that it's not something that will dominate

In my stream work, after years of working for total protection of the banks, I've been working more lately for the fish. The most stable channel in terms of erosion control is not the most ideal condition for fish. They need pools and diversity in the stream channel. I've started leaving certain structures in the channels, carefully placed, to cause downcutting, thus creating little pools that loosen up the gravel so the fish can spawn, and also creating nice living places (rearing habitats). I also put in woody debris, which I originally considered an erosion threat and unsightly, to provide food and hiding places for the fish. The riparian, or streamside, vegetative cover is important for the stream and for the fish. I plant redwoods, alders, and willows below the flood high water level because these trees can thrive with wet feet and being knocked over and covered with silt.

When I moved to our land 18 years ago, our stream was totally trashed. The loggers used the stream bed as a haul road to take out logs because in those days there were no forest practice rules. The stream then ran underground a lot, under yards of fill and tons of debris. We moved the debris out and placed it above the high water level, removed fill and defined the channel where necessary by placement of rock. We planted exposed banks with grass and trees (alder, redwood, and fir), and protected existing trees with tubes. It is amazing how fast it came back. Now our creek is a beautiful little stream with mossy banks and trees and pools, and it can support fish. ■

## International Reforestation Suppliers

*This is the source Jack Monschke uses for his seedling-protection tubes (see preceding article); he also praises their great service. No chainsaws, although there is a saw-blade-on-a-rod gizmo for harvesting Christmas trees that's bound to find its way into some bad slasher movie. Besides forestry gear, if you need to set up a weather station, fight a fire, read a map or measure just about anything, this is a good place to find the tools.*
—Richard Nilsen
[Suggested by Jack Monschke]

### International Reforestation Suppliers

Catalog **free** from International Reforestation Suppliers, P. O. Box 5547, Eugene, OR 97405; 800/321-1037 (in OR: 345-0579).

### McLEOD TOOL

A fire line favorite, the McLeod Tool offers many real scalping benefits with its combination rake and hoe features. Excellent for scalping light debris and vegetation, duff, and heavily burned sites.
- Constructed of tempered 12 gauge steel.
- Hoe is 11" wide — 5" deep.
- Rake features 6 heavy duty tines — 3½" long.
- Handle length — 48".
- Sh. Weight: 5 lbs.

| ITEM DESCRIPTION | UNIT | PRICE |
|---|---|---|
| Hazel Hoe Scalper Complete w/Handle | | $ 31.05 |
| Hazel Hoe Blade Only | | 24.00 |
| Replacement Handle | | 7.05 |

### HAZEL HOE SCALPER
**Osborne Pattern Adze Hoe**

Like the McLeod, the Hazel Hoe is another double duty tool for use during both fire season and planting season. Designed for heavy grubbing, trenching, and scalping, the Hazel Hoe will cut easily and deeply into thick grassy planting sites.
- Made of special alloy tool steel.
- Electrically tempered.
- Blade width — 6¼".
- Blade length — 5¾".
- Handle length — 36".
- Sh. Weight: 5 lbs.

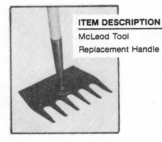

| ITEM DESCRIPTION | UNIT PRICE |
|---|---|
| McLeod Tool | $ 62.50 |
| Replacement Handle | 6.50 |

## Native Seed Foundation

*One of the real nitty-gritty issues in restoration work is genetic diversity within individual species. A rose ain't a rose ain't a rose, and neither is a salmon or a mountain hemlock tree. These fine-grained distinctions are overlooked when, say, Yellowstone burns up and gets re-seeded from airplanes hauling in seed grown two states away. Expediency and integrity are always a couple states apart anyway, so it is a real delight to find a small seed company trying to do something about this problem.*

*The Native Seed Foundation collects and sells seed of species native to the Pacific and Intermountain Northwest, and for their tree species at least, they tell you which national forest (or Canadian provincial park) the seed came from. Wholesale or retail, by the pound or by the ounce, here you can buy pinyon pine seed from Dixie National Forest in Utah, sub-alpine fir from Kootenai N.F. in Montana, or Douglas fir seed from five different locations. Additionally, they sell seed for over 30 native shrubs, and are also on the look-out for independent seed collectors they can buy from.*
—Richard Nilsen
[Suggested by Jack Monschke]

### Native Seed Foundation

Brochure **free**; Route W, Moyie Springs, ID 83845; 208/267-7938.

## Weed Wrench • Toadstool

*The endless contest between native plants and escaped exotic species is a big part of many restoration projects. Here are two tools developed by and for people doing this backbreaking work. Weed Wrench is a plant extractor, with pincer jaws that grab hold at ground level while a long handle levers out the roots. They come in four sizes and are currently being used on everything from banana puka on Molokai to Brazilian pepper in Florida.*

*Toadstool is an outdoor work seat, a modern version of the old one-legged milking stool. Your feet (or, on a steep hillside, your knees) form the other two legs of the tripod. It pivots like a swivel chair and makes crouching endurable.*
—Richard Nilsen

### Weed Wrench
**$50-$140** plus shipping
### Toadstool
**$19** postpaid
Both from New Tribe, 3435 Army Street #330, San Francisco, CA 94110; 415/647-0430

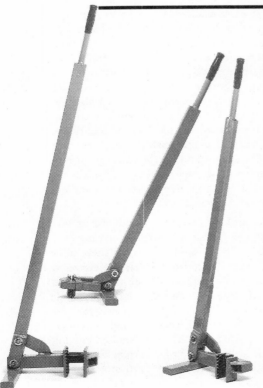

Root seedling

← slight bulge is root collar

hold seedling in hole with collar at ground level

pull out head and plunge in further back

## The Earth Manual

*Before environmental restoration, there was something called the back-to-the-land movement. This book, a product of that era, is proving itself venerable. It is still the best book for introducing beginners to the small-scale techniques of caring for land that are at the heart of restoration work.* —Richard Nilsen

●

Between well-trimmed suburban lawns and the vast regions of mountain wilderness, there are millions of patches of land that are semiwild. They may be wood lots, small forests, parks, a farm's "back forty," or even an unattended corner of a big back yard — land invaded by civilization but far from conquered. This book is about how to take care of such land: how to stop its erosion, heal its scars, cure its injured

### The Earth Manual
(How to Work on Wild Land Without Taming It)
Malcolm Margolin, 1975; 237 pp.

**$8.95** ($12.95 postpaid) from Heyday Books, P. O. Box 9145, Berkeley, CA 94709; 415/549-3564 (or Whole Earth Access).

trees, increase its wildlife, restock it with shrubs and wildflowers, and otherwise work with (rather than against) the wildness of the land.

## Bringing Back the Bush

*This nifty little book is about the art of pulling weeds. It is by two sisters who spent their lives perfecting a technique in the parklands of Sydney, Australia's Lower North Shore. Their method is selective and deft, all aimed at keeping exotic weeds from invading native ecosystems. You might call it the Zen of weeding.*
—Richard Nilsen
[Suggested by Michael Jack]

knife at an angle of about 45 degrees to the surface. This will bring the tip of the blade underneath the root.

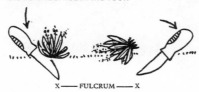

X —— FULCRUM —— X

push forward

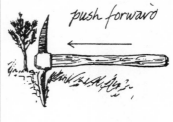

### Bringing Back the Bush
Joan Bradley, 1988; 111 pp.

**Out of print.** Formerly published by Weldon International, 371 Eastern Valley Way, Willoughby, NSW 2068, Australia

●

We began to understand that there might be another way to fight the invaders. Given half a chance, the bush would fight back on its own behalf.

●

Our weeding techniques minimise soil disturbance, muscular effort and working time. . . . You will not get results unless you follow our three principles meticulously.
• Work outwards from good areas to bad
• Make minimal disturbance
• Do not overclear
In every sort of plant community, wet or dry, rich soil or poor, rainforest, tall timber or heath, these principles remain the same.

●

Sawing straight through roots with a knife using an up and down motion seems to come naturally to many people but it is quite hard work, it takes a long time and therefore disturbs more soil than it should.

This is where the *thrust and lever* technique comes into its own by reducing the job to two easy motions. . . . Thrust about three-quarters of the blade into the soil, a little to the side of the root, with the back of the

## Adopting A Stream

*This excellent handbook grew out of the work of the Adopt-A-Stream Foundation in the Pacific Northwest, but its lessons apply to any stream or river that ever had fish in it. It details an effective method for involving elementary-school children in habitat restoration. They adopt sections of local streams and care for both the water and the fish that live there — in Washington state it is salmon returning from the ocean to breed.*
—Richard Nilsen

●

"The students have been incredibly enthusiastic about the Connelly Creek project from the beginning," says

**Storm drain stencils remind citizens to protect streams.**

Scherrer. "Not only have they learned about aquatic and wildlife biology, but they understand so much about how people and wildlife interact in the watershed. They now have a strong feeling of stewardship toward the fish, the stream, and the entire watershed."

### Adopting A Stream
Steve Yates
1988; 116 pp.

**$9.95**
($12.95 postpaid) from:
University of Washington Press
P. O. Box 50096
Seattle, WA 98145-5096
800/441-4115
or Whole Earth Access

## Stream Care Guide

*The task: to educate stream-side property owners and the general public about waterways and riparian habitat in two north-coast California counties. The solution: this brief and excellent pamphlet, done by a local private non-profit organization.* —Richard Nilsen
*[Suggested by Judith Goldsmith]*

•

Don't throw brush clippings or any other material into a stream channel for high water to "carry away". The brush may create an unwanted debris jam downstream on your neighbor's property. Streams are not garbage dumps!

Natural debris in the creek is needed for food and shelter for fish and wildlife. If flooding and bank erosion are not a problem, please leave the stream alone. "Cleaning" the stream usually hurts rather than helps fish and wildlife.

**(Right) An abundant variety of native plants along a stream creates good riparian habitat.**

### Stream Care Guide
Nancy Reichard, 1988; 14 pp.

**$1** postpaid from Natural Resources Division/Redwood Community Action Agency, 904 G Street, Eureka, CA 95501; 707/445-0881.

**T**HERE hasn't been a lot published yet on environmental restoration, especially for the general reader. It's a relatively new field, and more books are bound to come. For now the two best books, **Bioengineering for Land Reclamation and Conservation** (see Bibliography) and **The Earth Manual** (see preceding page), are each more than a decade old.

Before we are inundated by publishers trying to cover this subject, a suggestion — instead of broad-scale books, how about good, locally produced pamphlets from each bioregion. Some restoration techniques deserve a book treatment, but much of the nitty-gritty that allows this work to succeed involves specific local knowledge. Think of it as bioregional niche marketing, an opportunity for inexpensive publications from public service agencies or private regional presses. Here are three examples — of a 14-, a 60-, and a 96-page version of this kind of effort. —Richard Nilsen

David J. Cross

---

## Groundwork

*Too much of coastal California has been washing into the sea for a good chunk of this century. What logging didn't set loose the cows did. This handsome handbook is the product of one of America's more than 3,000 resource conservation districts. With a locally elected board, but no power to make or enforce laws, RCDs function as the glue between governmental bureaucracies and private landowners in cooperative efforts to conserve soil and water. Superb line drawings of techniques and a succinct text make this pamphlet one to emulate.* —Richard Nilsen

•

Willows are an effective and inexpensive way to armor active headcuts and gully banks.

Be sure to plant the willows right-side up. One almost foolproof method is to point the planting end of the sprig with an axe right after it is cut from the tree.

Willows spread easily — usually an advantage; but in some cases when an open channel is needed to carry stormflows, this can be a nuisance.

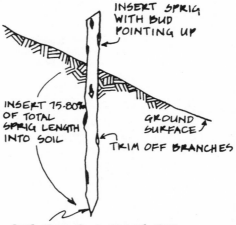

INSERT SPRIG WITH BUD POINTING UP

INSERT 75-80% OF TOTAL SPRIG LENGTH INTO SOIL

GROUND SURFACE

TRIM OFF BRANCHES

CUT END TO A POINT FOR EASIER INSTALLATION

**Willow sprig installation.**

### Groundwork
Liza Prunuske, 1987; 60 pp.

**$1.25** postpaid from Marin County Resource Conservation District, P. O. Box 219, Point Reyes Station, CA 94956; 415/663-1231.

# Stream Enhancement Guide

*Salmon habitat restoration is occurring around the entire Pacific Rim. This teaching manual from British Columbia's Salmonid Enhancement Program is ten years old and out of print, but the information it contains has a much wider audience than just one Canadian province. Aimed at both the public and government fisheries workers, it has some of the best before-and-after photos of any restoration booklet I've seen. At heart it is a manual of stream design, aimed at demonstrating solutions to salmon/human cohabitation. It deserves reissue as a watershed classic.* —Richard Nilsen

### Stream Enhancement Guide

Ministry of Environment, British Columbia, Canada, 1980; 96 pp. Out of print.

Concrete rectangular culvert with offset baffles bolted to the floor. The baffles were made of treated timber. Without them water velocities would be too high and the depth too shallow during fish migration.

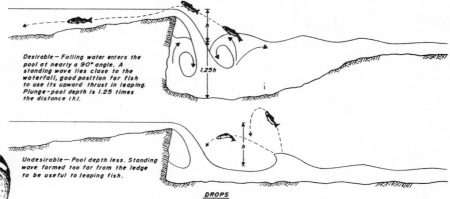

Desirable — Falling water enters the pool at nearly a 90° angle. A standing wave lies close to the waterfall, good position for fish to use its upward thrust in leaping. Plunge-pool depth is 1.25 times the distance (h).

Undesirable — Pool depth less. Standing wave formed too far from the ledge to be useful to leaping fish.

DROPS

# Attracting Backyard Wildlife

*This Canadian book, useful anywhere, gives careful consideration to little critters and little details. There is plenty of information on birds and small animals, but nice coverage of lizards and insects too. And for all, there are excellent diagrams of building housing and feeding stations.* —Richard Nilsen

•

Two ideas for improving backyards for toads are providing daytime retreats and hiding places, and a low power light source to attract night insects. Toad holes can be constructed in many imaginative ways, with soft sandy soil at the bottom being the key. . . .

Toad lights may be a light designed to illuminate a foot path. If the light is placed near a border between a garden or rockery and a lawn area, toads will have some cover while waiting for their meal to arrive.

### Attracting Backyard Wildlife

Bill Merilees, 1989; 159 pp.

**$10.95** ($13.95 postpaid) from Voyageur Press, 123 N. 2nd Street, Stillwater, MN 55082; 612/430-2210 (or Whole Earth Access).

# Wildlife Habitat Enhancement Council

*Corporations control lots of land, and often have little idea what's happening on it (think of the miles of power-line right-of-ways stretching across the landscape). This two-year-old organization is a cooperative venture between some very large U.S. corporations and some very mainline conservation groups. The goals appear to be restoring landholdings to environmental usefulness, making the public happy, and polishing tarnished images a bit. Let's see, that's win, win, win. If you live next to some BIG neighbors, these folks might be able to help you meet them.* —Richard Nilsen

### Wildlife Habitat Enhancement Council

Membership **$100**
1010 Wayne Avenue/Suite 1240, Silver Spring, MD 20904; 301/588-8994

•

*Case Study:* Delaware, Maryland and Virginia

*Company:* Delmarva Power & Light, Vienna Power Plant

*Habitat:* The Nanticoke River, a Chesapeake Bay tributary.

*Project:* Delmarva Power constructed a $50,000 brooding pond on its Vienna Power Plant grounds to help save the dwindling striped bass population in the Chesapeake Bay and its tributaries. Since 1985, Delmarva Power has released more than 40,000 striped bass fingerlings or rockfish into the Nanticoke River in Maryland.

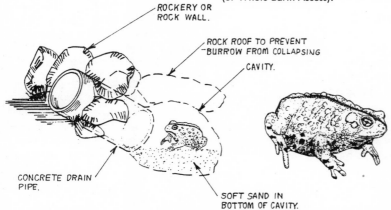

ROCKERY OR ROCK WALL.

ROCK ROOF TO PREVENT BURROW FROM COLLAPSING

CAVITY.

CONCRETE DRAIN PIPE.

SOFT SAND IN BOTTOM OF CAVITY.

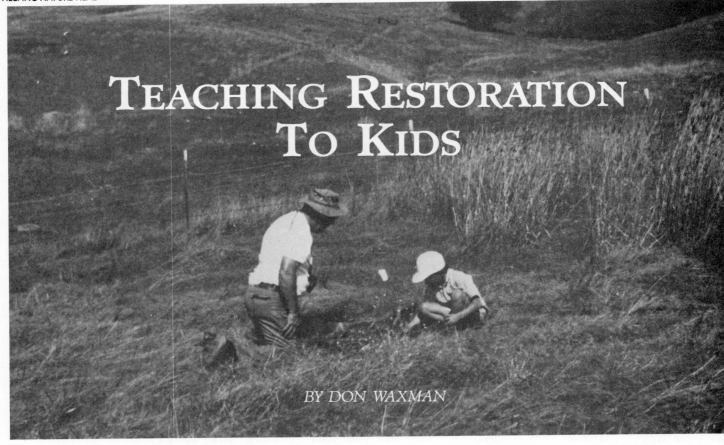

# TEACHING RESTORATION TO KIDS

*BY DON WAXMAN*

*Strategies for biological diversity are undergoing a dramatic conceptual shift. Nineteenth-century visions of untouched wilderness have given way successively, as has the landscape of North America itself, to preservation of contained parks and forests ("sanctuaries"), stewardship and, ultimately, intensive management of natural areas. In fact, "managed natural area" is no longer considered an oxymoron, evidence that we have come to accept (in many cases reluctantly) the role of human intervention in the perpetuation of naturally occurring systems. And although it may be difficult to forego a vision of "nature" apart from humanity, it is both practically and philosophically necessary to do so now. Restoration thus offers the brightest future — and perhaps the only future we would want to live in — for restoration ecology is simply the most advanced expression of the emerging ethic and may allow us to heal some of the damage that is being done to the planet.*
*—Don Falk, Center for Plant Conservation*

**I**N CONTRAST to the gloomy environmental predictions and scenarios that have characterized ecological thinking in the past, restoration ecology is an enterprise with the potential for optimism. We have no other choice but to attempt restoration ecology — it is an essential ongoing learning process intricately linked with our survival on this planet. As such, it provides a relevant and exciting context for educational programs.

Ecological awareness is attainable in many ways. What is most necessary is nature. One needs to go to nature to acquire a sense of the interrelatedness of all things. The ultimate experiences in nature are identification experiences; observing that natural objects and events are interrelated, and seeing oneself as an essential part of this interrelatedness. Through these experiences we realize that the interests of nature are our own interests. This is the heart of the new ecological consciousness. How can our young people acquire this kind of ecological consciousness when most of the time, most of them are completely out of touch with natural areas and ecological processes? Although today's world offers young people many choices and opportunities, their experience with natural areas and ecological processes is very limited.

I am establishing nature-study projects at several schools in my region. The goal of the program is to create tangible environmental resources on school sites, in the form of restoration projects where possible: interpretive native-plant display gardens in small spaces, and ecological landscapes in larger spaces. Using existing environmental-education curriculum materials, these plantings will provide hands-on experience by allowing students to actively participate in environmental problem-solving.

If you are interested in initiating restoration projects or native-plant projects on school sites, first form a project committee consisting of teachers, students, staff, and parents at each interested school. The committees will plan, establish, develop, and support restoration projects at their respective schools. Help each participating school

*Don Waxman runs an environmental design and planning business in Petaluma, California. He is acting chairperson of a nonprofit group, Petaluma Tree People, that organizes community tree-planting events on private and public properties. He can be contacted at 2805 Eastman Lane, Petaluma, CA 94952; 707/778-8529.*
*—Richard Nilsen*

develop an initial curriculum library. Connect your project with the projects of other local environmental organizations that have related kinds of concerns and agenda. Share resources, materials, and fundraising strategies with them.

Develop a landscape approach that makes the best sense given the school site and available resources. Teachers and staff are often unaware of unused and neglected land at their schools. Many of these areas are just never noticed by teachers. Those who do notice them usually do not recognize their ecological significance, or their value as potential outdoor classrooms. In most cases, a simple walk around the school grounds with teachers and staff to point out these unused resources is all it takes to generate enthusiasm for a native-plant project. Some schools already have many existing natural features which can be used in environmental-education activities.

A great educational exercise for starting a restoration project is to do a site survey and evaluation of these resources. This could include vegetation mapping, soil analysis, wildlife inventories, hydrological assessments, and listing past and current impacts of nearby land-use practices and developments. Similar study of local, relatively undisturbed natural areas should be conducted concurrently for contrast, as well as for ideas on how the school site might be ecologically restored or improved.

If the school site doesn't offer any significantly large areas for restoration projects or ecological landscaping, smaller areas can be developed using native plants to create interesting interpretive displays. For example, try displaying some of the endangered plants from your region to introduce the notions of rarity, endangerment, and the conservation of biodiversity; display dominant species of typical plant communities of your area to convey the plant-community concept; display plants according to families to study botanical taxonomy; display plants used by Native Americans and early settlers in your area — ethnobotany; display plants with unique adaptations — biogeography and evolution; display plants that attract and depend upon certain wildlife — coevolution, life cycles and symbiosis.

Many teachers think that they lack an adequate background for teaching a nature-based curriculum. Excellent environmental-education curricula have been developed and tested (see reviews); there is no need for teachers to think that they have to develop their own. Many of these materials have been developed for particular grade levels. Even without books, teachers need not think they are incapable of teaching environmental education. Restoration ecology projects demand a multifaceted approach, and teachers will be surprised to find out that much of their existing knowledge can be significantly applied. Those who know least about environmental education, but who are willing to learn along with their students, stand a good chance of becoming excellent outdoor educators.

An outdoor-education specialist might be employed to conduct inservice workshops with a teaching staff. Here are some suggested workshop topics:

• explore the various ways we experience nature and wilderness (or wild things — plants, landscapes, animals, etc.)

• define ecological restoration and conservation biology. Describe the ethical foundations for implementing restoratión projects. How do processes of valuation regarding nature and/or wild things influence our ethics? What is the conservation ethic? How can it be taught?

• develop techniques to assess student attitudes toward natural areas and ecological processes. Collect data on

Students and teachers plant native species to create a nature trail at Cherry Valley Elementary School in Petaluma, CA.

Photos by Don Waxman

student experiences with/in nature. Create methods for interpreting the data.

• survey and analyze local environmental problems and issues. Coordinate appropriate school-based activities that attempt to solve these problems. Integrate existing environmental-education lessons and activities into this effort.

Public-school teachers and staff are incredibly busy. In most cases, the staff doesn't have enough time to adequately complete programs that have already been started. Approach school administrators with a great deal of patience, and with as much understanding of their situation as is possible.

Don't expect an outdoor nature-study area to be funded with state-allocated funds for education — public schools are usually broke. Many of our schools simply do not have enough money to do what they are mandated to do. They are short of staff and materials, and there is usually a large wish list of things that need to be done. You will need to raise funds another way. Usually, parent-teacher associations raise funds throughout the year for different projects; local businesses, clubs, municipal departments, and other organizations will either provide funds or donate space, materials, and services for fundraising events. Another possibility is to find a nonprofit organization that is willing to adopt your program, and to help you write proposals for funds to foundations, private corporations, and government agencies.

Do not be stymied by bureaucracy. Some schools and school districts will have a lot of problems with any changes you propose for public-school properties. Risk management has become a major concern in our age of lawsuits. School and district administrators have to be especially careful. Be patient. ■

## Project Wild

*Project Wild is an interdisciplinary, environmental and conservation education program emphasizing wildlife. There are three curriculum guides — elementary, secondary, and one called* **Aquatic** *for grades K-12. Project Wild activities are organized around a conceptual framework comprising these major themes: awareness and appreciation of wildlife, human values and wildlife, wildlife and ecological systems, wildlife conservation, cultural and social interaction with wildlife, wildlife issues and trends, and responsible human actions regarding wildlife and ecological systems.*

*Each Project Wild activity includes objectives, method, background for the teacher, materials needed, procedures, evaluation suggestions, recommended grade level, school subject area(s), skills, duration, group size, setting, concept, and key vocabulary. The guides include a glossary and several cross-referenced indexes (lists of activities by grade level, skills acquired, topics covered, etc.) to help teachers incorporate wildlife-related concepts into their daily teaching strategy.*

*Project Wild is a joint project of the Western Regional Environmental Education Council and the Western Association of Fish and Wildlife Agencies.*
*—Don Waxman*

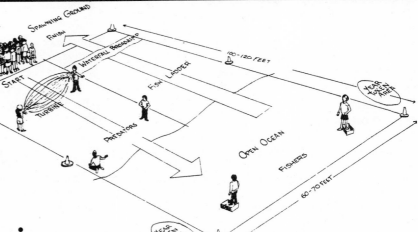

**HOOKS AND LADDERS**
*Method:* Students simulate Pacific salmon and the hazards faced by salmon in an activity portraying the life cycle of these aquatic creatures.

*Age:* Grades 3-9

*Materials:* Large playing area (100 feet x 50 feet); about 500 feet of rope, string, or six traffic cones for marking boundaries (masking tape may be used if area is indoors); two cardboard boxes; 100 tokens (3 x 5 cards, poker chips, etc.).

*Procedure:* . . . This is a physically-involving activity! Set up a playing field as shown in the diagram, including spawning grounds, downstream, upstream, and ocean. . . .

Assign roles to each of the students. Some will be salmon, others will be potential hazards to the salmon. . . .

Begin the activity with all the salmon in their spawning ground. The salmon then start their journey downstream. The first major hazard is the turbines at the dam. . . .

At most dams there are escape weirs to guide migrating salmon past the turbines. The student salmon **cannot go around the** jump rope swingers, but they **can slip under** the swingers' arms if they do not get touched while doing so. A salmon dies if it is hit by the turbine (jump rope). . . . NOTE: Any salmon that "dies" at any time in this activity must immediately become part of the fish ladder. The student is no longer a fish. . . .

Once in the open ocean, the salmon can be caught by fishing boats. The salmon must move back and forth across the ocean area in order to gather four tokens. Each token represents one year of growth. Once each fish has four tokens (four years' growth) that fish can begin migration upstream. . . .
*—Aquatic*

**Project Wild** activity guides are available free of charge to participants in Project Wild workshops. Workshops are open to educators and youth-group leaders at nominal charge. For more information, contact Project Wild, P. O. Box 18060, Boulder, CO 80308-8060; 303/444-2390.

### Manure, Meadows and Milkshakes

Eric Jorgensen, Trout Black, and Mary Hallesy, 1986; 132 pp.

**$9.95** ($11.45 postpaid) from The Trust for Hidden Villa, 26870 Moody Road, Los Altos Hills, CA 94022; 415/941-6119 (or Whole Earth Access).

## Manure, Meadows and Milkshakes

*This is a book full of wonderful games and activities that are every bit as creative as its title. The lessons effectively teach children the principles of ecology through hands-on exercises. One fifth-grader who participated in* **Manure, Meadows and Milkshakes** *activities said, "You have to learn about nature in person. You have to taste it, feel it, see it, and hear it."* —Don Waxman

•

Cinquains are a wonderful way to combine feelings and facts about our world into a poetic image. Easy to learn, easy to do! Here's how:

First line: one word, giving title
Second line: two words, describing title
Third line: three words, expressing an action
Fourth line: four words, expressing a feeling
Fifth line: one word, a synonym for the title

•

*Ranger Game:* Play this on a narrow trail when your children are pushing and shoving to be first. Number everyone off and start up the trail. No "cuts". Place yourself third in line. (You can talk to the two ahead and also keep track of the gang behind.) The first child is the ranger. When you or the children spot something of interest have your ranger stand by it and point it out to everyone as they walk by. "Ranger Dan or Ann" then becomes the caboose at the end of the line. Make sure everyone has a chance to be a ranger.

## Urban Forestry for Children

*An excellent set of activities and lessons about trees. It has a large section in the back that lists resources to be used in conjunction with the lessons. This list includes descriptions of each resource, and details about where they can be acquired. Most are free to teachers, or are very inexpensive. Lessons include "What is a Tree," "How Trees Grow," "The Seasons and Cycles," "Building a Tree," "Tree Identification," "Importance of Trees," and "Understanding and Caring for Trees."* —Don Waxman

### Urban Forestry for Children

Teachers may obtain this book free from Cooperative Agricultural Extension, University of California, 4145 Branch Center Road, Sacramento, CA 95827; 916/366-2013.

•

The Cottonwood (*Populus fremontii*) is mentioned more often than any other vegetation in the account of early exploration of the West. They were signs of water in arid lands and the sweet inner bark was food for horses. On January 6, 1844, John Fremont wrote: "Taking Kit Carson with me I made a thorough exploration of the neighboring valleys and found in a ravine in the bordering mountains a good camping place where there was water in the springs and a sufficient quantity of grass for the night. Overshadowing the springs were some trees of the sweet cottonwood which after a long interval of absence, we again saw with pleasure, regarding them as harbingers of better country." Because explorer, John Fremont often mentioned cottonwoods in his writings, botanists named the tree after him.

Cottonwoods were very important to the California Indians and pioneers. Indians used the wood for making drums. (Arizona Indians still use the wood for making their Kachina dolls.) The Indians believed the cottonwoods were beings of higher intelligence and that misfortune would follow if due respect were not paid to them.
—*Trees: Urban Forestry for Children*

## Project Learning Tree

*Like Project Wild, Project Learning Tree is an interdisciplinary, environmentally focused supplementary curriculum program. It is written by teachers and designed for integration into the regular curriculum at all grade levels. It focuses a bit more on trees and forests than does Project Wild, revealing the bias of the American Forest Foundation and the American Forest Council, which sponsor and administer the program in conjunction with the Western Regional Environmental Education Council. Project Learning Tree uses the forest as a "window" into the natural world, helping young people gain an awareness and knowledge of the world around them, as well as of their place in it.* —Don Waxman

**Project Learning Tree Guides** are available free to participants of Project Learning Tree workshops. For information, contact The American Forest Council, 1250 Connecticut Avenue NW, Washington, DC 20036.

## More for kids

*There are many other excellent environmental-education materials. Some very noteworthy materials can be acquired from* **Ranger Rick's NatureScope**, *National Wildlife Federation, 1412 16th Street NW, Washington, DC 20036-2266. Another excellent group of materials are the learning modules put out by* **Outdoor Biology Instructional Strategies**, *which can be obtained from the Lawrence Hall of Science, University of California, Berkeley, CA 94720.* —DW

# A Survey of Publications Covering Restoration   *BY DAVE EGAN*

*ENVIRONMENTAL restoration news is scattered far and wide, often in tiny newsletters, or magazines that cover it occasionally. Dave Egan scans these publications and writes up abstracts of what's interesting for **Restoration & Management Notes** (p. 76), where he is assistant editor. Here are his choices of the best publications. Like the subjects they cover, many of them are fragile — tiny budgets, no advertising revenues, few subscribers — and worthy of your support.   —Richard Nilsen*

## American Midland Naturalist

*One of the best experimental science journals available to the restoration community. Its articles are well-researched and -written, not overly laden with technical jargon or unruly statistical data. Recent examples include "A 12-year study of vegetation and mammal succession on a reconstructed tallgrass prairie in Iowa," "A 20-year study of the effect of prescribed burning on **Andropogon scoparius** in Connecticut," "Response of **Hamamelis virginiana** to canopy gaps." Definitely worth trying to obtain either as a subscriber or through your local college library.*

•

The explosive eruption of Mount St. Helens on 18 May 1980 resulted in a large-scale destruction of montane/subalpine forest and meadow habitats and caused the local extirpation of many small mammal species. At least 14 species of small mammals survived the initial eruption and concomitant habitat alterations. Individuals of many "residual" species survived the volcanic blast by virtue of their fossorial habitats; being in underground burrows at

### American Midland Naturalist
Robert P. McIntosh, Editor

**$50**/year (4 issues) from American Midland Naturalist, University of Notre Dame, Notre Dame, IN 46556.

the time of the early morning eruption shielded them from the rapidly moving hot gases and tephra. Although the blast virtually annihilated all exposed plants and arthropods in nearby areas, the existence of underground plant parts and invertebrates, combined with a limited "rain" of recolonizing arthropods onto the disturbed areas, provided an adequate food supply for some of the surviving animals. Localized aerial seeding by government agencies (as part of a reclamation effort) augmented some food resources.

## BioScience

*While fairly technical in its approach and subject material, this journal still provides restorationists with some interesting reading, particularly in the areas of genetics and biological diversity. Articles from recent issues include "Blueprint for conserving plant diversity," "Shredders and riparian vegetation," and "Integrated conservation strategy for Hawaiian forest birds."*

### BioScience
Julie Ann Miller, Editor

**$42**/year (11 issues) from AIBS, 730 11th Street NW, Washington, DC 20001-4584; 202/628-1500.

•

Geese on the west coast of Hudson Bay are turning salt marshes into mudflats. This environmental devastation may foreshadow scenes to come in the predicted global climate change. . . . Increased winter precipitation, thus deeper snow accumulation and greater cloud cover, would delay snow melting. During the last decade, areas along the shores of Hudson Bay have experienced late spring thaws — a weather pattern resembling some of the climatic predictions.

The northward spring migration of the lesser snow goose follows the melting snow. . . . In a study area at La Pérouse Bay . . . as many as 100,000 snow geese have remained for several weeks in spring, instead of stopping only for a day or two.

This migration delay has upset the balance that allows nondestructive grazing of grass and sedges. . . . In an hour one grubbing goose can clear a square meter of land that may then become dried mud in summer.

## Coast and Ocean

*This journal focuses on California coastline, although the information is applicable to other states as well. Recently funded dune, beach, and littoral wetland restoration projects are featured. By placing restoration within the context of a multiple-use resource, **C&O** provides a pragmatic look at the balancing act between human use and resource management. Recent articles include "Restoration standards: how to guarantee environmental protection," "Watershed restoration: an idea whose time has come — again," "How public access affects wetlands," and "Stream restoration: the healing touch."*

### Coast and Ocean
Rasa Gustaitis, Editor

**Free** (4 issues/year) from Coast and Ocean/State Coastal Conservancy, 1330 Broadway, Suite 1100, Oakland, CA 94612; 415/464-1015

•

The concept of mitigation may be seen as a form of plea bargaining in which a developer who causes environmental damage receives a "lighter sentence" by promising to perform environmental penance. Some mitigations are reasonably straightforward. Dedications of access or conservation easements, for instance, are clear cut, easy to verify, and their benefits are felt almost immediately. Restoration, however, is a special kind of mitigation. It is a long-term, experimental, sometimes expensive undertaking which often lacks clearly definable goals and whose success can never be fully guaranteed. This absence of certainty is often at the root of controversy surrounding restoration projects.

Some biologists involved in development-linked restoration projects have been surprised by the intensity of opposition they have encountered: they fail to understand that restoration is much more than a biological activity. It is also a political, legal, economic, and in some respects a philosophical enterprise.

## Canadian Journal of Botany

*Restorationists often rely on quantitative, scientific data to give some measure of predictability to their projects. This multi-faceted journal is an example of a publication where such information is often available. Its research-oriented articles do, however, require a background in statistical interpretation in order to be profitable. Topics from recent issues include "Relationship between mycorrhizal dependence and competitive ability of two tallgrass prairie grasses," "Postfire lichen — spruce woodland recovery at the limit of the boreal forest in northern Quebec," "Studies of mycorrhizal association in Harvard Forest, Massachusetts," and "Population density and height distribution of **Corylus Cornuta** in undisturbed forests of Minnesota, 1965-1984."*

•

The study of mushrooms in North America is faced with many difficulties. Mushroom taxonomy, although armed with sophisticated technology, is approximately 100 years behind that for vascular plants. There is great debate over generic concepts and names, many new species are still being described from temperate areas of the world, and species concepts themselves are often vague, just as they were for vascular plants 100 years ago. Within the past 20 years the literature aimed at the amateur mycologist has overtaken the scientific literature in many respects. Field guides give information on fungal distributions and other data not documented in the scientific literature. Precise information

### Canadian Journal of Botany

Iain E. P. Taylor, Editor

**$145**/year (12 issues); sample $17 from Research Journals/Subscription Office, National Research Council of Canada, Ottowa, Ontario, Canada K1A 0R6

on the distribution of mushrooms is difficult to determine from the literature because of the mixture of useful data with misinformation based on misdeterminations, differing species concepts, or the misapplication of names. There is even uncertainty about the ranges of some of our commonest species! . . .

The end result of this discrepancy between common knowledge and scientifically documented data is that we can only make very broad and vague statements regarding the overall distribution of this common macrofungus in Canada or North America.

Traditionally mycologists have hesitated to emphasize distributional patterns of macromycetes based on existing collections. Vast areas of North America are unexplored mycologically.

## Conservation Biology

*This relatively new, smart-looking journal carries very cogent articles about the international preservation and conservation of animal species, especially those which are threatened or endangered. Habitat and ecosystem restoration are discussed as a strategy for accomplishing this larger goal. Recent issues have had some restoration-oriented articles, including "Restoring island ecosystems: the potential of parasites to control introduced mammals," "Vegetation dynamics (succession and climax) in relation to plant community management," and "Consequences and costs of conservation corridors."*
*[Suggested by Dave Foreman]*

### Conservation Biology

David Ehrenfeld, Editor

**$39.50**/year (4 issues; includes membership) from Journals, Blackwell Scientific Publications, Inc., 3 Cambridge Center, Cambridge, MA 02142; 617/225-0401

•

Numerous mechanisms have existed over the past five centuries to move perhaps thousands of species of shallow-water marine organisms around the world in innumerable patterns. These mechanisms include the movement of species by ocean-going vessels — as fouling and boring organisms, in "dry" ballast (sand and rock) and in "wet" ballast (seawater) — as well as the equally unintentional movement of species sequestered with commercial fisheries products. . . .

Ballast water (not bilge water) is pumped into a vessel to alter the trim and stability of an empty or partially loaded ship. The plankton in the water pumped in includes organisms that spend their entire lives in the water column (the holoplankton, such as many copepods) and the larval stages of bottom-dwelling (benthic) species (the meroplankton). . . .

We have found that almost all major and most minor marine phyla survive such voyages and often arrive in vast quantities: over 200 species of living zooplankton and phytoplankton have been collected in the ballast water of 70 cargo ships arriving from Japan in Coos Bay, Oregon.

## Endangered Species Update

*In response to federal legislation, the U.S. Fish and Wildlife Service has developed thousands of recovery and/or reintroduction plans for endangered plant and animal species. They are discussed in the service's **Endangered Species Technical Bulletin**. Budget cuts in the Reagan years limited distribution, so the **Bulletin**'s contents are now reprinted in **Endangered Species Update** along with longer articles, opinion and book reviews. The total package provides an interesting look at one form of restoration — the restocking of individual species into their original environment.*

•

The photos of appealing red wolf pups in newspapers around the nation signal better times for this endangered mammal. Intensified efforts to breed red wolves produced 13 successful litters this past spring, increasing the population from less than 85 animals to nearly 120. The more than 30 new pups nationwide are excellent public envoys for the wolf and the key to its recovery.

The red wolf *(Canis rufus)* is native to the southeastern states where it was nearly extirpated as a feared and unwanted predator. By the early 1970s, the species had been reduced to a remnant population on the Anahuac National Wildlife Refuge and private lands along the Gulf Coasts of Texas and Louisiana. Researchers found the wolves there endangered by continued conflict with humans, parasites, and genetic swamping from hybridization with the coyote.

### Endangered Species Update

Joel Heinen and Alice Clarke, Editors

**$23**/year (10 issues) from the School of Natural Resources, University of Michigan, Ann Arbor, MI 48109-1115; 313/763-3243

## Fremontia

*This journal, like most native-plant society publications, seeks to educate the public on the need to appreciate and preserve their region's native plant populations. An intelligent, horticulture/genetics bent surfaces not only in its articles about rare-plant propagation, but also in probes regarding the long-range effects of restorative plantings amongst native plant populations. Good information on exotic species control is typically available, as well. Recent articles include "Restoration: Disneyland or a native ecosystem?" and "Can native flora survive prescribed burns?". Interesting reading no matter what your bioregion, and one of the best examples of what an organization might do to encourage others to value their local landscape and native flora.*

### Fremontia
Phyllis M. Faber, Editor

**$18**/year (4 issues) from California Native Plant Society, 909 12th Street, Suite 116, Sacramento, CA 95814; 916/477-2677

●

The Santa Ana River woolly-star in the Phlox family is one of California's rarest plants. It is now restricted to an area of about eight square miles along the Santa Ana River, north of the city of Redlands. The plight of the woolly-star was acknowledged in the fall of 1987 when it, along with another Santa Ana River species, the slender-horned spineflower, was listed as endangered by the federal government. Construction of the proposed Seven Oaks Dam near where the Santa Ana River emerges from the San Bernardino Mountains will . . . eliminate annual flooding and over a few decades all woolly-star habitat will disappear and the subspecies will become extinct.

**Monitoring young Santa Ana woolly-star plants growing in an open sandy habitat.**

## National Wetlands Newsletter

*For those interested in the legal and policy aspects of wetland restoration, this newsletter is definitely worth reviewing. Wetland mitigation, 404 permitting, "no net loss," and other wetland issues are discussed in guest columns. Standard features include updates on federal legislative action, pertinent legal suits, as well as to-the-point book reviews and a detailed events calendar. I always look forward to this engaging publication.*

●

The Coalition for the Conservation of Aquatic Habitat — otherwise known as FISH — is a rather unlikely alliance. . . . After agreeing that they had been unable to wield enough force *individually* to effect the preservation of fisheries habitat, the coalitions members decided that "it would be more fun to fight over *more* fish to

### National Wetlands Newsletter
Nicole Veilleux, Editor

**$48**/year (6 issues) from Environmental Law Institute, 1616 P Street NW, Suite 200, Washington, DC 20036; 202/328-5150

harvest than *less* fish." And thus, the FISH Coalition was born.

*FISH, 5410 Grosvenor Lane, Suite 110, Bethesda, MD 20814-2199; (301) 897-8616.*

---

## In the Courts

**United States v. Pozsgai**, Cr. 88-00450 (E.D. Pa. July 13, 1989). After filling a 14-acre tract of land in Pennsylvania despite a temporary restraining order and a court order to cease and desist, the defendant underwent a jury trial and was convicted for filling wetlands in violation of §404 of the Federal Water Pollution Control Act. The defendant was sentenced to three years in jail and ordered to pay a $202,000 fine and restore the damaged wetlands.

---

## Natural Areas Journal

*This publication provides reports from natural areas about rare species management and monitoring as part of the overall goal of ecosystem preservation. Although it is not NAJ's major emphasis, restoration has recently received more attention (and thus more journal space) as an important tool for the land manager. It is well written, with a good bibliography of recent publications about restoration and management. Articles from previous issues include "Structure and composition of low elevation old-growth forests in research natural areas of southeast Alaska," "Managing prairie remnants for insect diversity," and "Design of a long-term ecological monitoring program for Channel Islands National Park, California."*

●

The natural areas movement is not a new phenomenon. By 1917 the Ecological Society of America had already formed a Committee for the Preservation of Natural Areas. As many recognize, Victor Ernest Shelford at the University of Illinois was instrumental in organizing concern for natural areas. By 1926, with people like Henry Chandler Cowles at Chicago, Charles Christopher Adams in New York, Robert Wolcott in Nebraska, Forrest Shreve in the Sonoran Desert, Asa Orrin Weese at Oklahoma, and others, he produced a landmark volume, *Naturalist's Guide to the Americas* (Shelford 1926). In my opinion, nothing even remotely close to that work has been done since. The book is a treasure-trove of information, much of it still applicable, even though many of the areas that were discussed have long since vanished.

### Natural Areas Journal
Greg F. Iffrig, Editor

**$25**/year (4 issues) from Natural Areas Association, 320 S. 3rd Street, Rockford, IL 61104; 815/964-6666.

*Beaver dams such as this one at Mount Desert Island in Acadia NP are important factors in establishing habitat for a variety of wildlife including river otter. Beaver-created or modified wetlands support an abundance of fishes and frogs, major items in the otter's diet, as well as providing isolation and safe resting places.*

## Park Science

*While the National Park Service is the recipient of criticism from various circles, it nevertheless maintains a solid commitment to scientific, restoration-oriented activities. Their in-house newsletter, **Park Science**, documents those activities along with myriad other NPS projects. Previous issues have included articles entitled "Non-native mountain goat management undertaken at Olympia National Park," "Endangered species restoration at Gulf Islands Seashore," and "Galapagos botanical management provides perspective for Hawaii."*

### Park Science
Jean Mathews, Editor

**Free** (4 issues/year) from Park Science, 4150 Southwest Fairhaven Drive, Corvallis, OR 97333

## Prairie/Plains Journal

*A true grassroots adventure, this newsletter provides insights into ongoing restoration efforts on the native prairie, riverine, and other unique native habitats of the Nebraska landscape. Artistry, botany, and history are exquisitely combined to produce a delightful creation — a bonding of nature and humans through restoration activities.*

### Prairie/Plains Journal

**$10**/year (2 issues) from Prairie Plains Institute, 1307 L Street, Aurora, NE 68818; 402/694-5535

•

Mormon Island Crane Meadows [on the Platte River south of Grand Island] contains a diversity of habitat types and associated flora and fauna, many of which are unique. Over 500,000 sandhill cranes stage in the Platte River Valley for a month each spring, 5-9 million ducks and geese use this area for migration, breeding, and feeding, and four federally threatened or endangered bird species are dependent on the Platte Valley.

## Wildflower

*This journal is very easy to read, yet contains insightful horticultural information for both lay and professional audiences. Its articles are about native wildflowers, their propagation and use in home landscaping as well as in revegetation projects. While the intent is national in focus, many pieces deal with vegetation native to the southern and southwestern United States. Articles from a recent issue include "Task force recommends: wildflower policy for Minnesota," "Wild sunflowers: heritage and resource," and "Desert botanical garden research: owl-clover and commercial mixes."*

### Wildflower
John E. Averett, Editor

**$25**/year (2 issues) from National Wildflower Research Center, 2600 FM 793 North, Austin, TX 78725; 512/929-3600

•

Many people are surprised to discover that the honeybee, *Apis mellifera*, is only one of approximately 3,500 species of bees in America north of Mexico. Nearly all of those are native species; however, the honeybee, a native of the Old World, was introduced long ago by European settlers. Bumblebees, leaf-cutter bees, mason bees, mining bees, oil bees, digger bees, carpenter bees — the diversity of native species is remarkable.

*Other interesting journals and newsletters also cross my desk which are more local or specialized in nature. If they are located in your vicinity or deal with a topic in which you have an interest, I urge you to contact them. The local newsletters can be especially helpful for putting you in touch with restorationists in your area.*

### Plant Conservation
**$35**/year (4 issues) from Plant Conservation, Missouri Botanical Garden, P. O. Box 299, St. Louis, MO 63166.

### Creek Currents
**$15**/year (4 issues) from Newsletter of the Urban Creeks Council, 2530 San Pablo Avenue, Berkeley, CA 94702; 415/540-6669.

### Missouri Prairie Journal
**$20**/year (4 issues) from The Missouri Prairie Foundation, P. O. Box 200, Columbia, MO 65205.

### The Palmetto
**$20**/year (4 issues) from Florida Native Plant Society, 2020 Red Gate Road, Orlando, FL 32818; 407/299-1472.

### Wetlands Research Update
**Free** (2 issues/year) from United States Environmental Protection Agency/Corvallis Environmental Research Laboratory, 200 Southwest 35th Street, Corvallis, OR 97333.

POISONING your neighbors' air or water is just not the neighborly thing to do. Nor, for that matter, is stockpiling so much deadly poison that the slip of a match or a tiny valve leak could wipe out the entire neighborhood.

Yet somehow, American businesses have forgotten their good manners. They have fallen back on what government regulators will let them get away with — regardless of whether a common person's sense of decency would be offended.

Part of the reason some businesses have gotten away with this rude behavior for so long is that, until now, the public hasn't had the information needed to recognize this offensive behavior. Now this has changed. The federal Community Right-to-Know law enacted in 1986 has made the public aware that as much as 22 billion pounds of toxic chemicals are emitted to our air, water and land each year. These are chemicals that cause cancer and birth defects, and which attack virtually every organ system in the human body. This new law enables citizens to pinpoint geographically thousands of companies that are using and storing toxic materials.

The Right-to-Know law discloses the identities and types

# TURNING INDUSTRIAL POLLUTERS INTO GOOD NEIGHBORS

## BY SANFORD J. LEWIS

*Sanford J. Lewis is an environmental attorney in Acton, Massachusetts, who represents community and environmental organizations, and consults to community groups who want to act as citizen regulators. His clients include the National Toxics Campaign (see p. 137). An earlier version of this article appeared in the journal New Solutions.* —Richard Nilsen

of emissions from the major toxics users and polluters. The information is readily available in the "Toxic Release Inventory" available through regional offices of the EPA. Or any citizen with a computer and modem can dial up this same data through a new on-line database operated by the National Library of Medicine (see p. 138).

Through these data disclosures, people are coming to realize that companies just down the block may be major "chemical abusers." One result is that America's businesses are being reschooled by their neighbors as to what it really means to be a good neighbor. With the Right-to-Know information in hand, it becomes painfully obvious that our environmental laws and agencies are far from adequate to protect communities from unneighborly behavior by local businesses.

If the government isn't protecting us, what's to be done? Making governments accountable is part of the picture. Citizens can pressure their governments to do a better job of enforcement and to clean up local offenders. But in this time of shrunken government resources, an alternative strategy that is closer to home, and often more effective, is taking hold.

Citizens are going directly to local businesses and challenging their prerogatives to poison the air or water, and to put lives at risk. Even if the offending company's behavior is perfectly legal — condoned by bureaucrats in the State House or the White House — it just isn't neighborly. These "citizen regulators" have concluded that the time has come to demand a change. Ordinary citizens are beginning to knock on the polluters' doors. When those doors open, company executives are discovering that their neighbors are armed with common sense, data and expertise, and an arsenal of persuasion strategies.

For example, in Berlin, New Jersey, a group of residents known as the Coalition Against Toxics began an investigation after fishkills in nearby lakes. Members wondered if chemical discharges by the nearby Dynasill Company might have caused the fishkills. Dynasill produces glass for high-tech applications, including laser and aerospace uses.

In May 1988, with the company's permission, the citizens conducted their own inspection of Dynasill, with an industrial hygienist whose services were provided by the National Toxics Campaign. While it became apparent that the company had not caused any fishkills, at least recently, a number of other hazards were identified. A report prepared by the hygienist after the tour made a number of recommendations for improving the facility's chemical safety. For instance, the report recommended that the company complete its diking around storage tanks containing silicon tetrachloride, which, when exposed to water, can create heat and hydrochloric acid. It recommended installing showers and eye-wash stations. And it suggested training employees to be a company fire brigade.

Within one month of receiving the inspection report, the company implemented all of the recommendations that the group had made. In light of the good sense of the recommendations, this was clearly the neighborly thing to do.

Good-neighbor strategies do not always achieve such easy victories or instantaneous results. For instance, citizen regulators in Quinsigamond Village, a densely populated area in Worcester, Massachusetts, engaged in a year-long fight with the Lewcott Company.

These citizens were outraged by years of odorous pollution from the company, which applies resin coating to fabric. Government regulators seemed complacent — more intent on protecting the company's economic well-being than on protecting local residents' health. Yet the residents believed that the plant's emissions were the source of many illnesses in their neighborhood.

At the outset, these Worcester citizens met in one another's living rooms, and pored over the available information on the company's chemical usage patterns. They formed a new organization, the Quinsigamond Village Health Awareness Group (QVHAG).

Soon they launched an all-out campaign, declaring to the company and the community that from then on Lewcott would have to answer directly to its neighbors. The growing membership of QVHAG picketed the company. They inspected the company together with the National Toxics Campaign, and published an inspection report. They held press conferences. They challenged the company's flammables-storage permit, and won severe restrictions on the permit's duration. They pressed the state environmental agency to assess penalties for the firm's pollution. They lined a main street of Worcester with signs expressing their anger with the company's odorous behavior, such as ''DON'T RUIN OUR SUMMER AGAIN'' and even ''LEWCOTT GET OUT''. They investigated the company's directors, and the environmental records of related companies.

As a result of this flurry of action, Lewcott executives reluctantly agreed to meet with the residents. A series of formal negotiating sessions followed. The residents negotiated with the company point by point.

Changes in the plant managers' thinking began to emerge. Company executives began to realize that given the processes that their plant was engaged in and the large amounts of flammable materials used, their facility was simply too close to people's homes to ever be considered a good neighbor. They concluded that the most economical and effective means of becoming a good neighbor was to consolidate their Worcester plant and another Lewcott facility under one roof, with a new, more effective set of pollution controls.

A surprising and precedent-setting agreement resulted. The company actually committed, in a legally binding contract with QVHAG members, to the relocation of all of their operations a few miles away over a period of a year and a half. For the interim period during which they would remain in the neighborhood, they agreed to allow the citizens a right to inspect the facility. They also agreed to enforcement of their commitments — through a binding arbitration clause and designated penalties for any violations. In return, the citizens agreed to take down their signs and to ease up on their multifaceted campaign against the company. A smooth transition resulted, and a local environmental war came to an end.

There are dozens of other examples of local good-neighbor campaigns. Citizens in various communities are demanding that neighboring companies:

• Study and reduce toxic chemical usage and waste generation;

• Provide technical assistance to residents for review of a firm's activities;

• Allow residents the right to inspect factories periodically;

Members of the Coalition Against Toxics of Berlin, New Jersey on an inspection tour of the Dynasill Company plant in their community.

Richard Youngstrom

• Establish a comprehensive chemical-accident prevention program utilizing the advanced measures that the citizens' own experts identify; or

• Grant the citizens a right, along with workers in the plant, to ongoing participation in a company health and safety committee handling management decisions about toxics.

These strategies are proving most effective in dealing with small- and medium-size corporations. They are also being employed at big companies like Chevron in Richmond, California, and Exxon in Baytown, Texas. Concessions from these mammoth polluters have been harder to win.

THE GENIUS of the citizen-regulator approach is that it places those with the biggest stake in safety at the front lines of the regulatory process. What citizens can see for themselves, through direct inspections with their own ex-

This masonry wall at the Dynasill plant in New Jersey is built to contain any leak or rupture of the tank inside. Containment walls like this one were added by the company after Coalition members recommended them.

Tammy McGinley

---

## A SHORT LIST OF RESOURCE ORGANIZATIONS FOR GRASSROOTS ENVIRONMENTAL ACTIVISTS

**CITIZENS CLEARINGHOUSE FOR HAZARDOUS WASTE**
P. O. Box 6806, Falls Church, VA 22040; 703/276-7070

*Assistance to grassroots action groups. Publishes newsletter ($25/4 issues yr.) containing the best concise roundup of victories and fights at the grassroots. The staff of CCHW are experts at cutting through techno-jargon and bureacratic excuses. Their message is a simple one: ORGANIZE! ORGANIZE! ORGANIZE!*

**CLEAN WATER ACTION**
1320 18th Street NW, Washington, DC 20036; 202/457-1286

*Providing technical and organizing assistance to groups involved in fighting incinerators, cleaning up dumps and protecting groundwater. Also leading providers of strategy and support to citizens' efforts to inform the public on electoral candidates' environmental positions.*

**ENVIRONMENTAL RESEARCH FOUNDATION**
P. O. Box 73700, Washington, DC 20056-3700; 202/328-1119

*Provides weekly newsletter and on-line database for grassroots activists on solid-waste and hazardous-waste issues. (See RACHEL, p. 139.)*

**GREENPEACE USA**
1436 U Street NW, Washington, DC 20009; 202/462-1177

*The largest national grassroots-based group fighting toxics and solid-waste abuses, and also to conserve and restore the natural environment. Hard-hitting organizing and direct-action campaigns. A major international force that is changing the way the world views environmental issues.*

**INSTITUTE FOR LOCAL SELF-RELIANCE**
2425 18th Street NW, Washington, DC 20009; 202/232-4108

*An excellent resource for local activists and local government on the economics and technical ins and outs of solid waste.*

**NATIONAL COALITION AGAINST THE MISUSE OF PESTICIDES (NCAMP)**
701 E Street NE, Washington, DC 20003; 202/543-5450

*Technical assistance to groups and individuals fighting abuses of pesticides.*

**NATIONAL TOXICS CAMPAIGN**
1168 Commonwealth Avenue, Boston, MA 02134; 617/232-0327

*One of the most ambitious and far-reaching grassroots action and support groups in the world. Currently has 15 offices across the country. (See review, next page.)*

**US PUBLIC INTEREST RESEARCH GROUP (PIRG)**
215 Pennsylvania Avenue SE, Washington, DC 20003; 202/546-9707

*A great source of information on toxics and solid-waste issues. Engaged in national and statewide legislative fights. Offices of affiliated PIRGs are located in numerous states.*

**WORK ON WASTE**
82 Judson Street, Canton, NY 13617

*Publishes the weekly newsletter on solid-waste issues for grassroots activists. The founder, Professor Paul Connett of St. Lawrence University, is one of the most eloquent speakers in the nation on the need to shift to a nonwasteful economy.*

—Sanford Lewis ■

pert consultants, can provide information far more reliable and extensive than what a government inspector might report to them.

Citizen regulators should not think that they need to become experts themselves to make this strategy work. In fact, in almost all of the campaigns for good-neighbor agreements, the neighbors who are pressing the companies are typical working people, not "experts" at all.

The Lewcott campaign in Massachusetts was led by Cindy Phillips, a high-school gym teacher. The members of QVHAG include a school-bus driver, a postman, a police officer, factory workers and many other long-time residents of the blue-collar Quinsigamond Village neighborhood. The good-neighbor strategy engages ordinary people in common-sense approaches to secure their rights to community safety, sound neighborhood etiquette, and environmental justice.

Experience has shown that organized local citizens have as much, if not more, ability than government agencies to move local corporations to act. If a local company does not willingly negotiate, citizen regulators must be prepared to wage a campaign to bring the company to the bargaining table, by organizing, picketing, legal strategies, or generating unfavorable media publicity.

Citizen regulators often turn to government regulators for assistance. But instead of filing complaints only with a single environmental agency, the citizens may deploy dozens of agencies in fields as remote as zoning, banking and insurance toward their goal of pressuring a company to negotiate.

At its best, the good-neighbor strategy can bring an end to human isolation and despair. In order to take on a local polluter, many citizens who have never considered themselves to be activists before need to alter their attitudes and lifestyles. They need to be empowered, much as the nonsmoker who for the first time gains the courage to tell the cigarette smoker, "Yes, I do mind if you smoke!" Some citizen regulators must tear themselves away from their TVs, and get out and meet other residents with similar concerns. In short, they have to breathe new life into the notions of neighborhood and community by engaging in old-style civic activism.

Citizen regulators also discover that the typical "us-versus-them" lines are not drawn as they might at first believe. Natural allies are often situated within the factory gates, where workers are exposed to the same toxic chemicals that the community is concerned about. As the Dynasill example shows, many safety measures to protect the community will also help protect workers. Citizens will typically benefit by working with, not against, the unions or other representatives of workers.

From the other side of the picture, the isolation and security that many corporate executives have enjoyed until now is also being brought to an end by this good-neighbor strategy. Many company executives have mistakenly come to believe that the law grants them an absolute right to expose their neighbors to deadly pollutants. The public's expanded right to know is truly a first step toward stripping away the armor that protects polluters from being called on the carpet by their neighbors.

It no longer suffices for a company to ask bureaucrats far off in Washington, D.C., or their state capital for a seal of approval to poison local air or water. Across America, citizens are increasingly saying, "Let us be the final judge as to whether you are a good neighbor." ∎

## National Toxics Campaign

The National Toxics Campaign provides leadership, coordination and ammunition to grassroots toxics activists across the nation. NTC and its research arm, the National Toxics Campaign Fund, also bring expert organizational and technical help to grassroots groups.

One of the early projects of NTC was to coordinate grassroots groups to press Congress for a strong nationwide hazardous-waste-site cleanup law. Working with people at hundreds of dumpsites across the country, they collected more than two million signatures in support of Superfund. They also orchestrated a "Superdrive for Superfund," which featured a caravan of trucks collecting water and soil samples from 200 toxic sites around the country. The samples were delivered to the Capitol steps to emphasize the need for stronger legislation.

The Superfund law was an important first step. While cleaning up the past sites is a major job, America has not yet accomplished the prevention measures that are needed to stop the poisoning of its people and environment. So NTC has forged ahead with a multifaceted effort to **prevent** toxic hazards in all of their forms:

▶ NTC is providing tools to grassroots activists who are demanding that local polluters provide the Right to Know, the Right to Inspect and the Right to Negotiate.

▶ NTC launched a campaign to persuade grocers to end the sale of fresh produce with cancer-causing pesticides by 1995. As of this writing, more than a thousand grocers had signed up, including five major food chains.

▶ NTC is circulating and providing support for model state policies to reduce the use of toxic chemicals and the production of hazardous wastes. Legislative versions of the policies have been enacted in two states, and filed in about ten others.

▶ NTC publishes an excellent newsletter, **Toxic Times.**

▶ The platform in the NTC report **Shadow on the Land** brought family-farm organizations and environmental and consumer activists (including Greenpeace and Ralph Nader) together for the first time around a joint strategy to curtail the use of dangerous farm chemicals **and** keep family farmers on the land.

▶ During 1989, the organization established the first public-interest environmental laboratory. The lab provides truly independent testing for citizen activists who want to know the chemical ingredients of water, air, soil, sludges, wastes and tissues.   —Sanford Lewis

**National Toxics Campaign**
1168 Commonwealth Avenue, Boston, MA 02134; 617/232-0327

Membership in NTC is $25 for individuals and $100 for organizations. Membership includes subscription to **Toxic Times.** Organizational membership includes discount prices on lab services and publications.

# Finding And Using Toxics Information

### BY PHILIP WEXLER

YOU take a walk around your neighborhood and notice the ever-present chemical plant with its stacks pumping foul-smelling fumes into the atmosphere and spilling green liquid into the river. You wonder: What toxic chemicals are being released in my community? Which companies are responsible? Where are they transferring their toxic waste? How can I find out about the health effects of these chemicals on my family? Can they cause cancer or interfere with my wife's capacity to bear children?

Answers to such questions and others may be found by consulting an array of computerized databases, collectively known as MEDLARS (Medical Literature Analysis and Retrieval System), available through the National Library of Medicine (NLM) in Bethesda, Maryland.

NLM's Toxicology Information Program has been providing publicly available data about toxic chemicals and their hazards for over twenty years. In this time, its computer files have expanded greatly in number, size, and scope. Whereas this information has always been and continues to be of great interest to scientists and researchers, a greater effort is being made to inform concerned citizens that they, too, can have access to this enormous store of information at a very reasonable cost. Computer files available through MEDLARS' innovative Toxicology Data Network (TOXNET) have been a boon to individuals wanting to find out more about hazardous chemicals.

The Toxic Chemicals Release Inventory (TRI) is one of the newest of these TOXNET files. This file contains data on the estimated releases of toxic chemicals to the air, water, or land, as well as amounts transferred to waste sites. In compliance with the law, industrial facilities around the country are required to report this information to the Environmental Protection Agency, which in turn provides it to NLM, the public access point for this information on computer. This file helps fulfill the U.S. Congress's wish to insure the public's right to know about environmental toxic chemicals.

TRI offers very flexible searching. You may use the file to help determine how many pounds of chlorine, for example, have been released to the air in your zip code, or in your county, city, or state, and what facility has released the greatest amount for any given year. Alternately, you might choose to search on a company name, say ABC Chemical Products, and determine what substances they are releasing in your area, or nationwide. Listed with each company is a public contact name and phone number, should you need more information about the release. TRI can also be used if you want to find out how many pounds of certain hazardous chemicals manufactured or used outside your state are subsequently shipped into your state for disposal. These are but a few of the types of questions TRI is designed to answer.

TRI's versatility is enhanced through a statistical feature which allows you to perform a variety of functions such as adding numbers and calculating means, medians, maximums and standard deviations. One might also perform ''ranging'' to determine, for instance, what states release more than one million pounds of acetone into the air. In the near future, a ''sort'' capability will be implemented, allowing you to sort your results alphabetically or numerically. You might, for example, choose to sort the records you have retrieved alphabetically by company name.

Every database has its limitations. As helpful as the TRI data is, it contains only rather specific information related to chemical releases and transfers to waste sites. It does not contain information on the health effects of, or human exposure to, these chemicals. However, users of TRI are automatically authorized access to the entire family of NLM files, many of which do list such information and thereby supplement TRI.

The Hazardous Substances Data Bank (HSDB), for example, covers chemical toxicity, as well as emergency handling procedures, environmental fate, human exposure, detection methods, and regulatory requirements. A large part of the data in this 4,200-chemical file has been peer-reviewed by expert toxicologists and other scientists. The Registry of Toxic Effects of Chemical Substances (RTECS) is another TOXNET file of some 100,000 chemicals, covering acute and chronic effects, skin and eye irritation, carcinogenicity, mutagenicity, and reproductive

*Here is a brief tour through a portion of the alphabet soup simmering in Bethesda, Maryland, at the National Library of Medicine, the largest bio-medical library in the world. The acronyms are daunting but the information stored here can be as specific as your own back yard, and is available via computer. This is partly a result of the 1984 Bhopal chemical disaster, which thus far has killed over 3,000 people. The U.S. Congress responded to the question ''Could THAT happen here?'' with the Emergency Planning and Community Right-to-Know Act in 1986, which not only established the Toxic Release Inventory described here, but mandated that the information be made available by computer through telecommunications.*

*Philip Wexler is a part of the soup, working in the Toxicology Information Program at Bethesda. He is the author of* Information Resources in Toxicology.

—Richard Nilsen

consequences. TOXNET also includes files specifically dedicated to the areas of carcinogenesis, mutagenesis, and teratogenesis. Outside of TOXNET, the NLM's TOXLINE group of databases contains references, often with summaries, to journal articles dealing with hazardous chemicals and other areas of toxicology and environmental health.

Individuals may request searches on TRI and other NLM files by contacting any one of thousands of health science libraries and information centers throughout the country. Information on an online search facility in your community may be obtained from one of the seven regional medical libraries listed below. In the future, it is hoped that more public libraries will be tied into the NLM system. If you prefer to search these files on your own personal computer, you may request an application form from the address below. The cost of TRI and most other files averages $25 per hour with reduced rates during non-peak hours. The required equipment includes a computer terminal or personal computer, a phone line, modem, and telecommunications package.

User guides and instructions are available on all NLM files. A sequence of user-friendly, menu-driven screens has been developed for TRI to allow the novice or occasional user without any computer background to search the file. This interface leads you step by step through a series of menus which ask you what kind of information you need. You may then display your results at your personal computer and printer or have the results printed in an "off-line" mode at NLM and mailed to you the next day. Similar menus are being designed for TOXNET's other files.

We encounter potentially toxic substances at home, school, the workplace, and in the general environment. Living in a totally pure, pollution-free environment is not possible, though steps are being taken to lessen our technological dependence on and exposure to hazardous chemicals. Meanwhile, there is an increasing body of information about such chemicals to help us make decisions about the nature and degree of the hazards they present and how to manage their use. The National Library of Medicine, especially through its TOXNET computer system, is an invaluable resource for this toxics information.

For more information about TRI, the TOXNET system, or other National Library of Medicine files, please contact:

**Toxicology Information Program**
National Library of Medicine
8600 Rockville Pike
Bethesda, MD 20894
301/496-6531

### REGIONAL MEDICAL LIBRARIES
**The New York Academy of Medicine**
*2 E. 103rd Street*
*New York, NY 10029*
*212/876-8763*

**University of Maryland Health Sciences Library**
*111 S. Greene Street*
*Baltimore, MD 21201*
*301/328-2855; 800/638-6093*

**University of Illinois at Chicago Library of the Health Sciences**
*P. O. Box 7509*
*Chicago, IL 60680*
*312/996-2464*

**University of Nebraska Medical Center Library**
*42nd and Dewey Avenue*
*Omaha, NE 68105-1065*
*402/559-4326; 800/MED-RML4*

**The University of Texas**
*Southwestern Medical Center at Dallas*
*5323 Harry Hines Boulevard*
*Dallas, TX 75235-9049*
*214/688-2085*

**University of Washington**
*Health Sciences Library and Information Center*
*Seattle, WA 98195*
*206/543-8262*

**University of California**
*Louise Darling Biomedical Library*
*10833 Le Conte Avenue*
*Los Angeles, CA 90024-1798*
*213/825-1200* ■

## RACHEL

*An on-line database known as the Remote Access Chemical Hazards Electronic Library, RACHEL provides an extremely useful source of information on the hazards of particular chemicals and technologies. You can enter the name of a particular chemical, and get hazard profiles from the U.S. Coast Guard, and more detailed profiles on some that were developed by the State of New Jersey. You can also search quickly through thousands of environmentally relevant newspaper articles that have been abstracted in a separate part of the system. Access to the system is free to citizen activists, and billed on an hourly basis to others.*

*"Rachel's Hazardous Waste News" states that its purpose is "providing news and resources to the movement for Environmental Justice." On two sides of one sheet of paper, this weekly newsletter provides more incisive technical, strategic, and policy information on the environment than a year's subscription to* **Time** *or* **Newsweek**. —Sanford Lewis

•

We humans dwarf nature. For example,

we inject twice as much arsenic into the atmosphere as nature does, seven times as much cadmium, and 17 times as much lead. Measured by the metals that nature moves into the oceans via river discharges, we humans mobilize (through mining) 13 times as much iron as nature does, 36 times as much phosphorus, and 110 times as much tin. Small wonder that we find ourselves destabilizing the planet's mineral cycles, disturbing weather patterns, distressing vast regions of water and vegetation. We are like a giant blind mechanic, hammering the planet with our technology. Yet we are dependent upon nature for everything we own and everything we are. It is past time to take stock, to cut back, to get smart.

There are two basic ways to control pollution: (1) contain it all ("zero discharge"); or (2) wait until someone is harmed *and can prove they have been harmed*, then control whatever caused the harm. This second way is being tried by the industrialized world today.

### RACHEL
Database **free**; for newsletter and other information, contact Environmental Research Foundation, P. O. Box 3541, Princeton, NJ 08543-3541; 609/683-0707

# What's Underfoot
## Computer Tools For Restoration

*BY HANK ROBERTS*

**The map is not the territory.**
—*Korzybski*, General Semantics

**But you've got to know the territory.**
—*Meredith Willson*, The Music Man

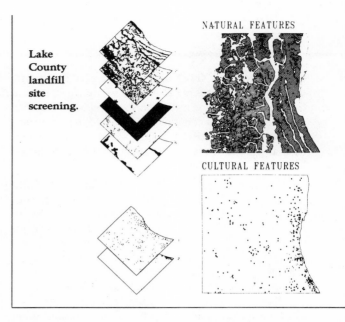

Lake County landfill site screening.

NATURAL FEATURES

CULTURAL FEATURES

A new and powerful computer tool is becoming available to environmental workers. Information that has been collected in database form — rows and columns of numbers and words — can be used to produce graphics, maps, and pictures. Geographic Information Systems (GISs) enable people to use those graphics as tools for handling and manipulating that collected information.

GISs begin with database systems. A map is produced from each database, and displayed using a graphics computer. Selecting a region on the map can help to define a query of the underlying database, and the query can be answered by a visible change on the map.

Currently, information in one database often can't be connected with that in another. Geographic information chosen by the database designer defines the map for that database: political boundaries, census tracts, zip codes, soil profiles, street layouts, easements, rights of way, pipes and wires, geology, age and condition of buildings, services used by residents, or the living features of the landscape.

A GIS brings together various databases by combining their maps. Each map is treated as a layer of information. It's possible to ask questions about a piece of the territory, and answer it from several databases which are represented on one combined map.

Most of the information that goes into a GIS is old information, and it will be no more accurate than its original users required it to be. To use these databases, their maps must be tied back to the earth — verified — to obtain what's called "ground truth." When each map has been verified the maps can then be traced or scanned.

The maps will not originally have been drawn to the same scale, or on the same coordinate system. To assemble a set of maps into a GIS, each map must be adjusted in a precise way. That's called "rubber-sheeting" — using the computer to stretch each map until reference points coincide when the maps are superimposed. Information from one database is then related to information in another database, and all the maps describe the same territory.

Accurate surveyors' maps are a good choice for a GIS because they are the legal description of the land, which any GIS is likely to have to conform to eventually. But someone planning a green belt would not need that level of accuracy and might not need a computer tool at all — as long as that project stayed independent of others' actions. A sketch on an existing paper map would suffice.

But multiple records, and multiple maps, lead to sadly familiar planning errors: streets repaved, then dug up to replace utility lines; sidewalk trees planted, then cut down for a long-planned widening of the street; streams restored by schoolchildren and then bulldozed by county engineers, as recently happened in California. As the cost of planning mistakes rises, the motivation to combine information and plan cooperatively may increase.

As GIS tools become available, they can be used to collect information from many sources. Once a GIS is running, it can produce maps for any database — and can be fed with any information a citizen might care to have a planner know about. Most of the information needed to produce a useful GIS is already available — somewhere.

This is the key point for citizens of a bioregion — a GIS for a region becomes a compelling tool for planning. A GIS will carry along with its information the documentation: Where did the information come from? How old is it? How accurate? That gives people the opportunity to question and correct these systems, and to include in them the facts that one gets by loving and living on the land and paying attention to it:

As a boy in Pennsylvania 40 years ago, I marveled at the swarms of warblers, vireos, flycatchers, thrushes, and other small colorful birds in our forests during May. Their songs were a continuous chorus. With practice, the ear could sort out and identify individual species within the chorus, just as individual instruments can be heard within an orchestra. Spring migration was one of the most spectacular events in the world of birds.

The swarms of birds, and the chorus, are not there anymore. By comparison with 40 years ago, spring has become almost silent. The chorus has dwindled to a few solos with gaps between; song identification is no longer much of a challenge.

DEVELOPMENT

HIGHWAY PROXIMITY

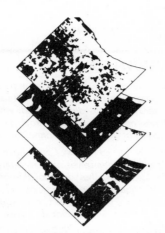

Using GIS, twelve mapped variables such as endangered species habitats, school and hospital sites, existing development, and suitability of highways to heavy truck traffic were summarized on four composites which allowed the Illinois Department of Energy and Natural Resources to determine areas of suitability for a proposed new landfill. ARC/INFO Maps

I would say that spring migration has declined by at least 80 percent since the 1940s.

My impressions are subjective. I have no data; I've even changed locations. But ornithologists whose memories go back 20 or more years all say the same thing: spring migrations have collapsed.

These birds travel high in the sky by night. The number of migrating birds in recent nighttime radar pictures has declined "precipitiously" — by more than 50 percent — from 1950s radar pictures.

Changes in bird populations sometimes tell us something . . . .

*—Dr. Charles F. Wurster, excerpted with permission from* Environmental Defense Fund Letter, *Oct. 1989*

A GIS need not be complex to be useful; it can begin as a simple system used to automate routine clerical work. The commonest need in local government is notification. Identifying every property owner who should be notified of a proposed zoning change, a tedious task when done by hand, can be done by marking the affected area on the GIS zoning map rather than by searching through the files. Upgrading the software later can add query features such as plan checking — pointing to a parcel on the map and calling up the relevant codes and requirements to confirm that it is acceptable for a proposed use.

The next generation of GIS programs will be able to use "distributed" software systems that live nowhere in particular and can be used from any telephone (such as Usenet). Using these tools, local groups of people will be able to share the environmental records they have collected, by combining their databases, allowing locally collected information to be fitted together into a picture of the world we share.

Within a couple of years such networks will be accessible to local restoration workers — and that's not much lead time to get your information ready. The database information is the core material, that which lasts, changing only to reflect changes in the physical world. New GIS software and hardware may be obtained frequently as these tools are improved.

As the cost of the tools decreases, the cost of labor involved in collecting the information will rise. This is where a GIS can be a tool for the eyes and ears and mind. Human memory is the best and most fragile record of what the world is like. Much of what we know about the natural world resides in aging human memories, scattered and anecdotal. That information can be collected using the simplest tools — handwritten notes will do for a start — and such local detail will improve the resolution of the Big Picture when one is made. Memories, notes, pictures and recordings can become the baseline data needed as restoration work proceeds. ∎

# The Sierran Biodiversity Project

The Sierran Biodiversity Project was initiated to develop long-range planning tools for the maintenance of biodiversity in California's Sierra Nevada mountains. We are establishing a geographical information system (GIS) for Sierran National Forest lands, using data from the Forest Service and other sources. The coverages being constructed include elevation, roads, hydrography, ownership and management boundaries, vegetation and soil typing, animal distribution, and past and prospective forest cutting. Using the composite maps, ancient forests will be identified and field-verified. The insights from conservation biology and landscape ecology will be applied to design a master biodiversity network and maintenance plan.

In the past, a mapping project of this scale would have involved hundreds of people and years of time — to produce a set of hard-copy maps that would be out of date upon completion. The real power of GIS technology is analogous to the power of a word processor in its phenomenal ease of map updating and editing. Additionally, the powerful numerical tools of the GIS permit rapid quantitative analysis of geographic features and their attributes.

*—Eric and Steve Beckwitt*
*P. O. Box 29*
*North San Juan, CA 95960*

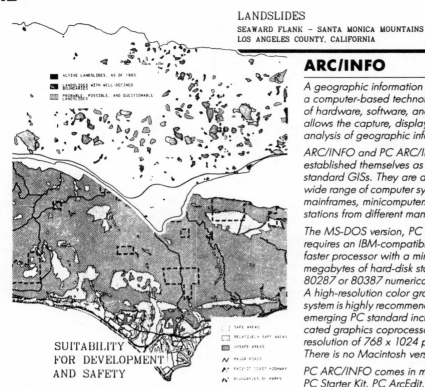

LANDSLIDES
SEAWARD FLANK - SANTA MONICA MOUNTAINS
LOS ANGELES COUNTY, CALIFORNIA

ACTIVE LANDSLIDES, AS OF 1983
LANDSLIDES WITH WELL-DEFINED BOUNDARIES
PROBABLE, POSSIBLE, AND QUESTIONABLE LANDSLIDES

SAFE AREAS
RELATIVELY SAFE AREAS
UNSAFE AREAS
MAJOR ROADS
PACIFIC COAST HIGHWAY
BOUNDARIES OF PARKS

SUITABILITY
FOR DEVELOPMENT
AND SAFETY

*One variable and a composite in a series of maps which can be used as a model for problem-solving and decision-making for planning purposes.*

## ARC/INFO

*A geographic information system (GIS) is a computer-based technology consisting of hardware, software, and data, which allows the capture, display, editing and analysis of geographic information.*

*ARC/INFO and PC ARC/INFO have established themselves as the industry-standard GISs. They are available on a wide range of computer systems including mainframes, minicomputers, and work-stations from different manufacturers.*

*The MS-DOS version, PC ARC/INFO, requires an IBM-compatible 80286 or faster processor with a minimum of 40 megabytes of hard-disk storage and 80287 or 80387 numerical co-processor. A high-resolution color graphic display system is highly recommended. The newly emerging PC standard includes a dedicated graphics coprocessor and screen resolution of 768 x 1024 pixels or better. There is no Macintosh version available.*

*PC ARC/INFO comes in modules — PC Starter Kit, PC ArcEdit, ArcPlot, PC Overlay, PC Data Conversion and PC Network. The fully installed program occupies about 29 megabytes of hard-disk storage.*

*The ARC system stores geographical feature locations as a set of coordinates and as relationships between connected or adjacent features. The relationships used to represent the connectivity of contiguous features are referred to as topology. The geographic features are digitally stored in computer files as layers or "coverages." Typical coverages assembled for a USGS standard 7½-minute quadrangle (1:24,000 scale) might include elevation, soil types, vegetation types, hydrography, or roads and ownership boundaries.*

*After learning the ARC/INFO software system, input of graphic and feature attributes is the largest task associated with building a GIS. Maps may be digitized or scanned into computer files and then imported into ARC coverages. Standard graphical output is to a plotter or to a Postscript- or Laserjet-compatible laser printer. Some dot-matrix printers are supported at lower graphic resolutions.*
*—Eric and Steve Beckwitt*

### ARC/INFO

For more information, contact: Marketing Dept., Environmental Systems Research Institute, Inc., 380 New York Street, Redlands, CA 92373; 714/793-2853.

## macGIS

*Want a GIS right away? The macGIS program begins with a scanned map or aerial photograph for the base picture. A grid is created and aligned precisely with the base map. Numerical information may be "painted" over this map by hand using the mouse, or generated using imported data from which macGIS paints a picture. Overlays assign paint patterns to represent numeric values for any mappable characteristic. When a new observation is available, it may be painted onto the data layer, and the underlying database is updated as that is done.*

*The macGIS User's Guide is a 200-page workbook, clearly written and well illustrated, showing its academic origins. The user is assumed to know how to use a database; reports are produced by arithmetic and algebraic op-*

*erations on existing data. What macGIS offers is the graphic interface, whereby a database query produces a new map. A complicated analysis can be recorded to produce a batch program which can redo the procedure automatically.*

*The novice learning to use macGIS can use the basic Mac Plus or SE models, although quite slowly, making this program the only GIS available at present that will run on inexpensive computers.*
*—Hank Roberts*

### macGIS

Version 1.0

**$300** (price lower to educational and nonprofit groups). Mac IIx, hard disk, and plotter recommended. David W. Hulse, Dept. of Landscape Architecture, School of Architecture and Allied Arts, University of Oregon, Eugene, OR 97403; 503/686-3634.

## Text Processors

*Text processors are small fast programs for producing ASCII (plain text) files. For getting information written down for later use, they're all you need to avoid retyping. Here are two good ones.*

*QEdit: Fast! Many files can be open in memory at the same time, allowing text to be easily cut and pasted between files. QEdit offers both pull-down menus with mouse support and control-key commands. Shareware and commercial versions are available.*

*Vantage: An earlier version of this handy program, under the name MacSink, was available as shareware. It has been a sturdy performer for several years. The commercial version under the name Vantage is well supported and comes with a good manual.* —Hank Roberts

### QEdit

Version 2.1 DOS

**$57.95** postpaid from SemWare, 4343 Shallowford Road/Suite C3, Marietta, GA 30062; 404/641-9002.

### Vantage

Version 1.5; Macintosh

**$99.95** from Preferred Publishers, Inc., 5100 Poplar Ave./Suite 617, Memphis, TN 38137; 901/683-3383.

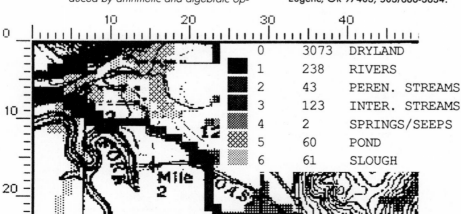

| | | |
|---|---|---|
| 0 | 3073 | DRYLAND |
| 1 | 238 | RIVERS |
| 2 | 43 | PEREN. STREAMS |
| 3 | 123 | INTER. STREAMS |
| 4 | 2 | SPRINGS/SEEPS |
| 5 | 60 | POND |
| 6 | 61 | SLOUGH |

Bird identification expert Beth Ann Gilroy examines a human artifact to determine if the feathers are from an endangered species. Identifications are based on both visual and microscopic comparisons against known feathers.

# A Crime Lab For Animals

*BY JUDITH GOLDSMITH*

OVER one hundred nations have now signed the Convention on International Trade in Endangered Species (CITES). As one of the signators, the United States pledges that no material made from or including parts of endangered animal species will be allowed to cross our borders or be sold within them.

The first National Fish and Wildlife Forensics Laboratory for enforcement of the Endangered Species Act and the CITES Treaty opened its doors in June 1989, in the quiet town of Ashland, Oregon. It is the first crime lab for animal species in the world.

Forensic scientist Ken Goddard, the lab's director, has been working since 1979 to make the lab a reality. Goddard was hired for the job because of his previous experience as director of a police crime lab. For the Fish and Wildlife Service to be secure about winning its battles in court, it needed a forensics lab to provide full evidentiary data to back up arrests and seizures.

Goddard writes mystery novels on the side. His sense of humor is a good thing, because the challenges of this work are enormous. As he points out, species of plants and animals are classified by certain groups of characteristics. For example, the differences between an endangered red wolf (now found only in remnant populations in Texas and Louisiana) and a legal coyote are usually based on the coyote's smaller size and lesser weight, its narrower snout and face, its less doglike appearance, smaller, less rounded ears, smaller feet, and black-tipped tail that droops low behind the hind legs. But, as Goddard says, "What if you only have a paw? Some teeth? Or worse, just spattered blood on a coat or shirt? What if the animal is indeed an endangered red wolf? Is it full-blooded or hybrid? It's legal to kill and sell hybrid wolves."

So the first challenge facing the forensics lab is developing a database of a whole new group of "species-defining characteristics," over and above what taxonomists use, that can be proven to be invariably indicative of a certain species of animal. Goddard describes the work that lab employee Dr. Ed Espinoza is doing on ivory, which is often seized as a carved art piece, resembling its original form only in color, texture, and hardness. "The first and easiest job is to prove that the material is real ivory, and not a synthetic look-alike. Beyond that, ivory from a (legal) warthog has to be differentiated from ivory from a hippo, walrus, sperm whale, or elephant. All of this scientists can already do, through visual, microscopic, and instrumental analysis techniques.

"Now comes the real challenge — proving that the ivory is from, say, an African or Asian elephant, which is illegal to kill, and not from a long-extinct mammoth or mastodon, which can be sold and worked without any penalty." With the help of an infrared spectrometer, Espinoza is able to make these identifications. The spectrometer is a non-destructive analytical tool (which allows the carving eventually to be donated to a museum); Espinoza is also working on identifying chemical structures in the crosshatching of ivory, a kind of species fingerprint.

Goddard showed me around the serology and morphology sections of the lab, where the challenges are just as high. In serology — identification through analysis of blood and tissue — a study of immunology may show that blood came from a member of the cat family. Protein electrophoresis could tell that the blood came from a snow leopard. But the lab also hopes to start using DNA analysis, which would actually prove that a blood sample came from a particular animal found in the back of a hunter's truck, mounted on a wall, or as dressed meat in a freezer.

The forensics lab is also amassing information from the only places that regularly deal with wild species of animals — zoos. In order to say that a particular fur or blood type comes from an endangered snow leopard, a sample from a snow leopard is needed for comparison. "I'm not about to go grabbing a handful of fur through the cage bars, or trying to poke the creature with a needle," Goddard jokes. Zoos regularly anesthetize live animals for medical reasons, and can then take documented blood, fur, or other samples which are preserved and catalogued for future use. Oliver Ryder of the San Diego Zoo has been

(Above) An assortment of Asian medicinal products containing ground rhinoceros horn. Rhinos are protected by the CITES treaty. Seized products like these can only be used as evidence in court if the protected materials are positively identified.

(Left) Forensic scientists Mary Jacque Mann and Ed Espinoza using a scanning electron microscope to identify a suspected piece of elephant ivory. Visual analyses, using a protractor to measure the distinctive angles of striations in ivory, are performed as well.

Lab technician Jim Kniffen using a bar-code scanner to log protected sea turtles into the storage repository. Typically, U.S. tourists bring these stuffed souvenirs home from Caribbean vacations, only to have them confiscated at customs by Fish and Wildlife inspectors. These turtles will eventually be distributed to schools and used for educational purposes.

Photos from National Fish and Wildlife Forensics Laboratory

saving fist-sized pieces of numerous animals in a cryogenic collection he hopes one day to use to recreate vanished species, and since he only needs thumb-size pieces, there has been discussion of donating some of his stock to the lab. Goddard is contacting many more zoos for help in obtaining data samples.

In other departments, carefully computerized inventory techniques are being developed to prove the chain of custody after a piece of evidence arrives at the lab. Human involvement with killings is proved by test-firing confiscated guns and analyzing human blood and fingerprints (after all, the animal may have died naturally, and have been found dead by the human in question). A scanning electron microscope analyzes gunshot residue, or pollen which might identify what part of the world a sample comes from. A mass spectrograph can discern one part sea turtle oil in 1,000 parts lanolin. In the warehouse, 50,000 to 60,000 items may be stored at any one time. "That number is kind of a wild guess," says Goddard. "New York or New Jersey could double our inventory in one afternoon simply by seizing a cargo ship."

Disposition of the evidence after cases are resolved will eventually become a problem. Goddard would like to hear from public or private agencies that could use seized evidence for educational displays about the dangers and problems of species destruction, such as the one at Denver's Stapleton Airport.

The new lab is currently budgeted for eighteen scientists in a total staff of thirty. That's adequate for the initial, standard-setting work, but will soon be grossly inadequate when scientists begin to start spending their time flying all over the world to testify at trials. Goddard estimates that the lab may eventually need a staff of 100 to keep up with the anticipated workload. "We're pioneers here. There are only eight states in the U.S. with their own wildlife crime labs, typically with only one or two people on staff. If each state could establish a crime lab adequate to cover its own cases, using the techniques we're developing, that would help a lot. If they could do just the one-hunter/one-animal cases, and leave us the complex cases, that would be a good first step. We'll be happy to share information." ∎

*Judith Goldsmith is a writer and graphic artist who does volunteer work with the Sierra Club on creek restoration projects in the San Francisco Bay Area.*

*The National Fish and Wildlife Forensics Laboratory can be contacted at 1490 E. Main Street, Ashland, OR 97520. Director Ken Goddard says his staff is happy to be living and working in southern Oregon. Forensic scientists normally have to go where the work is — to high-crime areas.*
                                                                                    —Richard Nilsen

# ORGANIZATIONS THAT HELP

The following lists are by no means complete. Much restoration work is regional and site specific. Three groups that may know more about projects in your area are your local chapters of the Audubon Society and The Nature Conservancy and your local native plant society. And don't forget one of the most valuable, if often overlooked and underfunded, resources available: the reference librarian at your local public library.

Page number citations refer to our reviews.

*Alliance for the Chesapeake Bay,* 6600 York Road, Baltimore, MD 21212. 301/377-6270.

*Auroville International USA,* P.O. Box 162489, Sacramento, CA 95816 (P.94).

*California Native Plant Society,* 909 12th Street, Suite 116, Sacramento, CA 95814. 916/477-2677 (P.132).

*California State Coastal Conservancy,* 1330 Broadway, Suite 1100, Oakland, CA 94612. 415/464-1015 (P.130).

*Center for US-USSR Initiatives,* 3268 Sacramento Street, San Francisco, CA 94115 (P.102).

*Citizens Clearinghouse For Hazardous Waste,* P.O. Box 6806, Falls Church, VA 22040. 703/276-7070 (P.136).

*Clean Water Action,* 1320 18th Street NW, Washington, DC 20036. 202/457-1286 (P.136).

*The Conservation Foundation,* 1250 24th Street NW, Suite 500, Washington, DC 20037. 202/293-4800.

*Ducks Unlimited,* 1 Waterfowl Way, Long Grove, IL 60047.

*Earth First!,* P.O. Box 5871, Tucson, AZ 85703. 602/622-1371.

*Earth Island Institute,* 300 Broadway, Suite 28, San Francisco, CA 94133. 415/788-3666.

*Environmental Defense Fund,* 257 Park Avenue South, New York, NY 10010. 212/505-2100.

*Environmental Law Institute,* 1616 P Street NW, Washington, DC 20036. 202/328-5150.

*Environmental Research Foundation,* P.O. Box 73700, Washington, DC 20056-3700. 202/328-1119 (P.136).

*Everglades Coalition,* Theresa Woody, 1201 N. Federal Hwy., Rm 250-H, N. Palm Beach, FL 33408 (P.62).

*Florida Native Plant Society,* 2020 Red Gate Road, Orlando, FL 32818. 407/299-1472 (P.133).

*Friends of the River,* Fort Mason Center, San Francisco, CA 94123. 415/771-0400.

*Global ReLeaf,* American Forestry Association, P.O. Box 2000, Washington, DC 20013. 202/667-3300 (P.72).

*Green Belt Movement,* P.O. Box 67545, Nairobi, Kenya.

*Greenpeace USA,* 1436 U Street NW, Washington, DC 20009. 202/462-1177 (P.136).

*Institute For Local Self-Reliance,* 2425 18th Street NW, Washington, DC 20009. 202/232-4108 (P.136).

*Izaak Walton League of America,* 1701 North Fort Myer Drive, Suite 1100, Arlington, VA 22209.

*Mattole Restoration Council,* Box 160, Petrolia, CA 95558 (P.51).

*Missouri Botanical Garden,* P.O. Box 299, St. Louis, MO 63166 (P.133).

*Missouri Prairie Foundation,* P.O. Box 200, Columbia, MO 65205 (P.133).

*National Audubon Society,* 801 Pennsylvania Avenue SE, Suite 301, Washington, DC 20003. 202/547-9009.

*National Fish and Wildlife Forensics Laboratory,* 1490 E. Main Street, Ashland, OR 97520 (P.145).

*National Toxics Campaign,* 1168 Commonwealth Avenue, Boston, MA 02134. 617/232-0327 (P.136).

*National Wildlife Federation,* 1412 16th Street NW, Washington, DC 20036-2266.

*National Coalition Against the Misuse of Pesticides (NCAMP),* 701 E Street SE, Washington, DC 20003. 202/543-5450 (P.136).

*Native Seed Foundation,* Star Route, Moyie Springs, ID 83845. 208/267-7938 (P.122).

*Natural Areas Association,* 320 S. 3rd Street, Rockford, IL 61104. 815/964-6666 (P.132).

*Natural Resources Defense Council,* 40 West 20th Street, New York, NY 10011. 212/727-2700.

*The Nature Conservancy,* 1815 North Lynn Street, Arlington, VA 22209. 703/841-5300.

*Northwest Coalition for Alternatives to Pesticides (NCAT),* P.O. Box 393, Eugene, OR 97440. 503/344-5044.

*Planet Drum Foundation,* P.O. Box 31251, San Francisco, CA 94131. 415/285-6556 (P.72).

*Prairie Plains Institute,* 1307 L Street, Aurora, NE 68818. 402/694-5535 (P.133).

*Project Learning Tree,* American Forest Council, 1250 Connecticut Avenue NW, Washington, DC 20036 (P.129).

*Project Wild,* P.O. Box 18060, Boulder, CO 80308-8060. 303/444-2390 (P.128).

*Rainforest Action Network,* 301 Broadway, Suite A, San Francisco, CA 94133. 415/398-4404.

*Restoring the Earth,* 1713 C Martin Luther King, Jr. Way, Berkeley, CA 94709. 415/843-2645 (P.6).

*Sierra Club,* 730 Polk Street, San Francisco, CA 94109. 415/776-2211.

*Sierra Club Legal Defense Fund,* 2044 Fillmore Street, San Francisco, CA 94115.

*Sierran Biodiversity Project,* Eric & Steve Beckwitt, P.O. Box 29, North San Juan, CA 95960 (P.141).

*Society for Ecological Restoration,* University of Wisconsin Arboretum, 1207 Seminole Highway, Madison, WI 53711. 608/263-7888 (P.76).

*Toxicology Information Program,* National Library of Medicine, 8600 Rockville Pike, Bethesda, MD 20894. 301/496-6531 (P.139).

*TreePeople,* 12601 Mulholland Drive, Beverly Hills, CA 90210. 818/769-2663 (P.45).

*Trust for Public Land,* 116 New Montgomery Street, 4th Floor, San Francisco, CA 94105. 415/495-4014.

*Urban Creeks Council,* 2530 San Pablo Avenue, Berkeley, CA 94702. 415/540-6669 (P.133).

*US Public Interest Research Group (PIRG),* 215 Pennsylvania Avenue SE, Washington, DC 20003. 202/546-9707 (P.136).

*Wilderness Society,* 1400 I Street NW, 10th floor, Washington, DC 20005. 202/842-3400.

*Wildlife Habitat Enhancement Council,* 1010 Wayne Avenue/Suite 1240, Silver Spring, MD 20904. 301/588-8994 (P.125).

*Work on Waste,* 82 Judson Street,

# BIBLIOGRAPHY

## Books

Ahrenhoerster, Robert and Wilson, Trelen. *Prairie Restoration for the Beginner.* Franklin, WI: Wehr Nature Center, 1988 (P.39).

Anderson, Walter Truett. *To Govern Evolution: Further Adventures of the Political Animal.* San Diego: Harcourt Brace Jovanovitch, 1987 (P.26).

Berg, Peter; Magilavy, Beryl; and Zuckerman, Seth. *A Green City Program.* San Francisco: Planet Drum Foundation, 1989 (P.72).

Berger, John J., Editor. *Environmental Restoration.* Covelo, CA: Island Press, 1990 (P.39).

Berger, John J. *Restoring the Earth.* New York: Random House, 1985.

Berry, Wendell. *The Unsettling of America: Culture and Agriculture.* San Francisco: Sierra Club Books, 1977.

Berry, Wendell. *What Are People For?* Berkeley: North Point Press, 1990.

Bradley, Joan. *Bringing Back the Bush.* Willoughby, NSW, Australia: Weldon Publishing, 1989 (P.123).

British Columbia Ministry of Environment. *Stream Enhancement Guide.* Victoria, BC, Canada, 1980 (P.125).

Cohen, Gary and O'Connor, John, Editors. *Fighting Toxics: A Manual for Protecting Your Family, Community and Workplace.* Covelo, CA: Island Press, 1990.

Dasmann, Raymond F. *Environmental Conservation.* New York: John Wiley and Sons, 1984.

Douglas, Marjory Stoneman. *The Everglades: River of Grass.* Englewood, FL: Pineapple Press, 1988.

Elliott, Jeff and Sayen, Jamie. *The Ecological Restoration of the Northern Appalachians.* North Stratford, NH: Preserve Appalachian Wilderness, 1990.

Giono, Jean. *The Man Who Planted Trees.* Post Mills, VT: Chelsea Green Publishing Co., 1985 (P.112).

Gridley, Karen, Editor. *Man of the Trees: Selected Writings of Richard St. Barbe Baker.* Willits, CA: Ecology Action, 1989 (P.112).

Hecht, Susanna and Cockburn, Alexander. *The Fate of the Forest.* New York: HarperCollins, 1990 (P.105).

Hunter, Christopher J. *Better Trout Habitat: A Guide to Stream Restoration and Management.* Covelo, CA: Island Press, 1990.

Jorgensen, Eric; Black, Trout; and Hallesy, Mary. *Manure, Meadows and Milkshakes.* Los Altos Hills, CA: The Trust for Hidden Villa, 1986 (P.129).

Kusler, Jon A. and Kentula, Mary E.. *Wetland Creation and Restoration: The Status of the Science.* Covelo, CA: Island Press, 1990.

Leopold, Aldo. *A Sand County Almanac.* New York: Oxford University Press, 1949.

Lipkis, Andy and Katie. *The Simple Act of Planting a Tree.* Los Angeles: Jeremy Tarcher, 1990 (P.45).

Margolin, Malcolm. *The Earth Manual: How to Work on Wild Land Without Taming It.* Berkeley: Heyday Books, 1985 (P.123).

Maser, Chris. *Forest Primeval.* San Francisco: Sierra Club Books, 1989 (P.58).

Maser, Chris; Tarrant, Robert F.; Trappe, James M.; and Franklin, Jerry F. *From The Forest to the Sea: A Story of Fallen Trees.* Washington, D.C.: U.S. Government Printing Office, 1988 (P.58).

Maser, Chris. *The Redesigned Forest.* San Pedro, CA: R. & E. Miles, 1988 (P.58).

McPhee, John. *The Control of Nature.* New York: Farrar, Straus & Giroux, 1989 (P.26).

Merilees, Bill. *Attracting Backyard Wildlife.* Stillwater, MN: Voyageur Press, 1989 (P.125).

Mills, Stephanie. *In Praise of Nature.* Covelo, CA: Island Press, 1990.

Misrach, Richard and Misrach, Miriam W. *Bravo 20: The Bombing of the American West.* Baltimore: Johns Hopkins University Press, 1990 (P.21).

Moll, Gary and Ebenreck, Sara. *Shading Our Cities.* Covelo, CA: Island Press, 1989 (P.72).

Mollison, Bill. *Permaculture: A Practical Guide for a Sustainable Future.* Covelo, CA: Island Press, 1990.

Newman, Peter; Neville, Simon; and Duxbury, Louise, Editors. *Case Studies in Environmental Hope.* Perth, Australia: E.P.A. Support Services, 1988 (P.97).

Pauly, Wayne R. *How to Manage Small Prairie Fires.* Madison, WI: Dane County Park Commission, 1985 (P.39).

Price, Robert. *Johnny Appleseed: Man and Myth.* Bloomington: Indiana University Press, 1954.

Prunuske, Liza. *Groundwork.* Point Reyes Station, CA: Marin County Resource Conservation District, 1987 (P.124).

Reichard, Nancy. *Stream Care Guide.* Eureka, CA: Redwood Community Action Agency, 1988 (P.124).

Rock, Harold W. *Prairie Propagation Handbook.* Franklin, WI: Wehr Nature Center, 1981 (P.39).

Schiechtl, Hugo. *Bioengineering for Land Reclamation & Conservation.* Edmonton: University of Alberta Press, 1980.

Shattil, Wendy and Rozinski, Bob. *When Nature Heals: The Greening of Rocky Mountain Arsenal.* Boulder, CO: Roberts Rinehart, 1990.

Soulé, Michael E. and Wilcox, Bruce A., Editors. *Conservation Biology: An Evolutionary–Ecological Perspective.* Sunderland, MA: Sinauer Associates, 1980.

Soulé, Michael E., Editor. *Conservation Biology: The Science of Scarcity and Diversity.* Sunderland, MA: Sinauer Associates, 1986.

Yates, Steve. *Adopting A Stream.* Seattle: University of Washington Press, 1988 (P.123).

## Periodicals

*American Midland Naturalist* (P.130).

*BioScience* (P. 130).

*Canadian Journal of Botany* (P.131).

*Coast & Ocean* (P.130).

*Conservation Biology* (P.131).

*Creek Currents* (P.133).

*Endangered Species Update* (P.131).

*Fremontia* (P.132).

*Missouri Prairie Journal* (P.133).

*National Wetlands Newsletter* (P.132).

*Natural Areas Journal* (P.132).

*Palmetto* (P.133).

*Park Science* (P.133).

*Plant Conservation* (P.133).

*Prairie/Plains Journal* (P.133).

*Restoration & Management Notes* (P.76).

*Wetlands Research Update* (P.133).

*Wildflower* (P. 133).

# ABOUT THE *WHOLE EARTH CATALOGS* AND *WHOLE EARTH REVIEW*

*Originally conceived by Stewart Brand in 1968, The Whole Earth Catalog* developed into a series of remarkable volumes of concentrated generalist information:

- *The Last Whole Earth Catalog* (1971)
- *The Whole Earth Epilog* (1974)
- *The Next Whole Earth Catalog* (1980)
- *The Whole Earth Software Catalog* (1984)
- *The Essential Whole Earth Catalog* (1986)
- *Signal* (1988)
- *The Fringes of Reason* (1989)
- *The Electronic Whole Earth Catalog* (on CD-ROM; 1989)
- *The Whole Earth Ecolog* (1990).

An ongoing quarterly magazine version of *The Whole Earth Catalog* was launched in 1974. Each issue of *Whole Earth Review* (originally *CoEvolution Quarterly*) is 144 pages of provocative articles and authoritative reviews of unusual books and useful tools for the hand, heart and mind, with an emphasis on acquiring skills and information. And, remarkably in the world of magazine publishing, *Whole Earth Review* carries no advertising (except for a couple of pages of subscribers' classifieds).

New subscriptions to *Whole Earth Review* are $20/year (four issues) for individuals, $28/year for institutions. Add $6 for Canadian and foreign surface mail; add $12 for airmail delivery anywhere in the world.

To order the *Ecolog* or *Whole Earth Review,* see the form on the facing page. For up-to-date information on all our available publications, drop us a note at our address.

## FINANCES

Whole Earth's projects are generated under the umbrella of Point Foundation, a nonprofit organization mandated to encourage educational projects and conjure up innovation. Open accounting having always been a Point modus operandi, here is a breakdown of how the money you money you spent on this book gets divvied up:

- 42% to Ten Speed Press;
- 42% to the bookseller;
- 8% to the book wholesaler;
- 8% to Whole Earth (includes payments to authors).

## HOW TO SUBMIT STUFF TO US

You are invited to submit comments, articles, reviews, photographs, artwork, and suggestions for possible use in our publications. We pay for what we use. For details, send SASE to Assistant Editor at our address.

**OUR ADDRESS:**
Whole Earth Catalog and Review, 27 Gate Five Road, Sausalito, CA 94965; 415/332-1716

## POST HOST-IES

Post Host-ies is a communication experiment wherein readers of Whole Earth publications participate by mail in "conferences" devoted to discussion of specific topics. Among the active groups, as this book goes to press, is one focusing on environmental restoration. For information, send SASE to "Post Host-ies" at our address.

## ORDERING BOOKS

"Access" is a *Whole Earth Catalog* tradition. The reviews in this book supply complete information for ordering from publishers and suppliers.

You have several alternatives in addition to ordering books directly from the publisher: you can ask for the book in your local bookstore or library, or you can order from Whole Earth Access Company (left).

## HOW TO ORDER FROM WHOLE EARTH ACCESS

Any book that has "or Whole Earth Access" under its ordering information may be ordered by mail or telephone from Whole Earth Access (WEA) in Berkeley, California. This company was inspired by the *Whole Earth Catalog* more than 20 years ago, but has always been a separate and independent company. We list WEA as a convenience to our readers, who may want to order from a single source instead of dealing with various publishers. To order from WEA:

1. List the titles and quantities of books you want. It is helpful to indicate the page numbers on which they appear in *Helping Nature Heal.* Start with the list (not the postpaid) price. Total the prices of the books. Add $3.50 postage for one book, $5 for two to five books, and 50¢ for each additional book (for delivery in California, add local sales tax).

2. Enclose payment in full via check or money order. VISA/MasterCard customers: print name from card, account number, expiration date, and sign your name. Credit-card orders may also be called in.

3. Include your street address. All orders are shipped UPS, which does not deliver to P. O. box numbers.

4. For foreign orders, shipping is $5 for the first two books and $1 for each additional book. Send check or money order in US dollars.

**Whole Earth Access**
**822 Anthony Street**
**Berkeley, CA 94710**
**800/845-2000; 415/845-3000**

# Like what you see in this book?

# Take a look at the next issue of *Whole Earth Review*.

## Free of charge.

Start a one-year subscription to *Whole Earth Review*. Take a look at a FREE trial issue. If you like it, you pay just $20 for three more issues (you get four in all). If you don't, just return our invoice marked "cancel," and you owe us nothing. Either way, you get to keep the trial issue.

Take advantage of this opportunity to try out *Whole Earth Review* at no risk.

*Whole Earth Review* guarantees to challenge your notions of what the world — or a magazine — should be. You'll find feature articles on the cutting-edge issues of the day: electronic democracy, access to political tools, virtual reality, women's wisdom, and much more.

Curious? Why not see for yourself? Fill out the form below and mail it back to us. We'll get you your FREE trial issue right away.

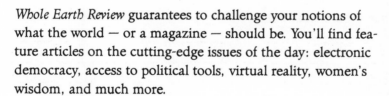

☐ **Yes, send me a FREE trial issue of *Whole Earth Review* to look at. If I like it, I'll pay just $20 for three more issues (four in all). If I don't like it, I'll just return your invoice, marked "cancel," and owe you nothing.**

☐ **Yes, send me ____ copies of the *Ecolog* for $17 each, postpaid. (Payment must accompany book orders.)**

Name _____

Address _____

City, State, Zip _____  PHNH

Mail this form to **Whole Earth Review**, 27 Gate Five Road, Sausalito, CA 94965.

FREE
MAGAZINE

# A ~~PENNY~~ FOR YOUR THOUGHTS

Dear Reader,

We hope you've enjoyed *Helping Nature Heal*. In the event that we update the book, we'd appreciate having your opinion of it — in particular how it might be improved. If you'll take a minute to give us a little feedback, we'll be happy to send you a recent issue of our magazine, *Whole Earth Review*, at no cost.

What did you like best about *Helping Nature Heal*?

_____

_____

_____

What did you like least about *Helping Nature Heal*?

_____

_____

_____

What wasn't in *Helping Nature Heal* that should have been?

_____

_____

_____

If this were a traditional questionnaire, we'd start asking about how many cars and VCRs you're going to buy this year — all for the benefit of our advertisers. But since *Whole Earth Review* doesn't carry any advertisements, we can skip that. All we'd like to get is an idea of who's talking to us, so please tell us your age, sex and occupation. Feel free to make any other comments.

Many thanks for helping.

Name _____

Address _____

_____ Age _____ M / F

Occupation _____     HNH

Send your valuable thoughts to:

**Helping Nature Heal/Whole Earth Review/27 Gate Five Road/Sausalito, CA 94965**

*PS: If you'd like to subscribe to* Whole Earth Review, *fill out this questionnaire and the order form on the reverse side of this page and we'll send you five issues for the price of four — so you'll still get a free issue.*

*PPS: If you'd like to be in touch with other individuals interested in environmental restoration, see page 152 for a description of POST HOST-IES — Whole Earth's round-robin conference-by-mail network.*

*If you're reluctant to cut out this page, send us a photocopy.*